Springer Business Cases

Springer Business Cases is a book series featuring the latest case studies in all areas of business, management, and finance, from around the world. The well-curated case collections in each of the books represent insights and lessons that can be used both in the classroom as well as in professional contexts. The books also place a focus on regional and topical diversity as well as encouraging alternative viewpoints which bring the knowledge forward. Both teaching cases as well as research cases are welcome.

Gitte Haar

Nordic Case Collection on Sustainability and Transition to a Circular Economy

 Springer

Gitte Haar
Center for Circular Economy
Hellerup, Denmark

ISSN 2662-5431   ISSN 2662-544X   (electronic)
Springer Business Cases
ISBN 978-3-031-78637-2   ISBN 978-3-031-78638-9   (eBook)
https://doi.org/10.1007/978-3-031-78638-9

This Springer imprint is published by the registered company Springer Nature Switzerland AG
The registered company address is: Gewerbestrasse 11, 6330 Cham, Switzerland

Preface

This Case Collection is a sequel to the textbooks by the same author to provide a review and a case collection of the Nordics and give insights into the need and implications of sustainability and the transition to a green and circular economy. The textbooks are:

- The Great Transition to Green and Circular Economy—Climate Nexus and Sustainability. 2024. Gitte Haar. Springer Nature.
- Rethink Economics and Business Models for Sustainability—Sustainable Leadership based on the Nordic Model. 2024. Gitte Haar. Springer Nature.

These two textbooks provide a scientific description of the state of the planet and give solutions on how to transform companies and society into a green and circular economy based on sustainable measures. These books also include concrete guidance for companies, organizations, and societies on how to implement the green transition. They also provides the basic understanding of the necessity for a new economy that operates within the planetary boundaries and communicate science in an understandable way. They explain the need for an overall transition and put companies and the business environment in the center of this transition. The tools presented in the textbooks are developed from many years of experience together with companies in the Nordics and are tested with these companies. The market conditions are changing rapidly these years becoming a business imperative. Regulation on circular economy and the need to document sustainable products based on life cycle analysis are rapidly being introduced, but also consumer demands are becoming more and more based on sustainable footprints.

In my work with companies, I meet many business leaders and politicians who still believe that sustainability is something that is done alongside traditional business. This is coming to an end and sustainability is now a business imperative and will be the largest change that companies will meet since the beginning of industrialization. For the past 15 years, I have worked with circular economy and green transition in companies, and during that time the world has changed a lot. The need for change in the way we live and run businesses is now clear to most people. Climate change has set in with rising temperatures, rapid melting of ice, extreme weather, forest fires, and floods. Human spread and overconsumption harm nature to such an extent that we gradually threaten our own habitats and livelihoods. In

2016, we were given the Sustainable Development Goals (SDGs) by the UN and companies have begun to adopt them as a communication platform. However, I still experience that leaders grope their way forward when working strategically with sustainability and the green transition.

Companies are central to the green and circular transition, and the transition is urgent. Actions and ambitious strategies need to be implemented. This is my aspiration to try to contribute by providing companies and others an understanding of why genuinely to embark on a green transition, but also how companies create a profitable business and anchor circular economy, climate neutrality, and the SDGs strategically into the core. I also want to give a picture of the need to change the economic models that shape the societies and the global linear value chains that built this world. We need to price access to nature and raw materials, and we need to understand that we long ago have passed the line for sustainable living. Now we must create the conditions at all levels to regenerate nature, the climate, and our economies and societies. We need to dare think differently, become innovative, and need to measure and account for impacts in completely new ways.

The big premeditator is to understand the entire value chain and the full footprint of the products we consume and how businesses operate in these complex supply chains. New business models must be developed, and information on the impacts of products throughout their life cycle must be provided in a traceable and transparent way. Above all, management must understand the complexity, barriers, and challenges of driving the great transition. The transition requires new non-financial ESG data to create traceability and transparency. Data that companies are not at all used to handle and that customers and consumers do not understand. The legislation for demanding these data is in place with the EU Reporting Regulation (CSRD and SFR) and the reporting requirements merging globally. Reporting of sustainability data (ESG) will be as important for companies and customers as the financial data.

I find that many business leaders are not aware of the changes that await them just around the corner. We already know the conjectures of the new market conditions and the new business models in the EU with the Green Deal Roadmap. Therefore, I would like to share my experiences and my knowledge about the green and circular transition from a business perspective, in the hope of accelerating the transition to a fair and sustainable planet. An agenda that is becoming urgent if we are to ensure a planet with a dignified life and prosperity for all. The transition to a green and circular economy also gives companies new business opportunities if they dare to rethink and innovative.

This Nordic Case collection and all the descriptions of the companies that have worked intensively with circular business models, sustainability, and the SDGs will hopefully inspire others to transform to a green and circular economy. In particular, the SMEs in Chap. 3 have shown the way and courage to innovate and create the strategic changes into the core of their businesses. Some of the SMEs considered frontrunners in Denmark and I have met and worked with in my work over the past 15 years are included in this Case Collection in Chap. 3. They have also contributed to the development of methods and tools described in the two books: *The Great Transition to a Green and Circular Economy* (Haar, The Great Transition to a Green and Circular Economy. Climate Nexus and Sustainability, 2024a) and *Rethink Economics and Business Models for Sustainability* (Haar, Rethink Economics and Business Models for Sustainability. Sustainable leadership based on the Nordic Model. 2024b). At the same time, the Nordics have several global corporations that are internationally noted for their work and communication on sustainability. 29 selected corporations are reviewed in Chap. 2.

The three books focus on economic and environmental sustainability, but also break with the traditional perception of corporate responsibility in society. They also give an overall view on how societies and organizations must change, and how we built our economic models and business models to be able to account for the impacts on the planet and on society. The books are intended as practical handbooks for companies and everyone who wants to be part of the great transition by ensuring new market conditions and new conditions in society and use the business and their competitiveness to create a sustainable future. With the books I hope to create a deeper understanding of the Climate Nexus, circular economy, climate change, and the SDGs than the general knowledge provided by media and hearsay. The books promote a holistic approach and try to bring an understanding of the complexity of sustainability. Sustainability and the green transition are the nerds' battleground.

I want to thank all the contributors to this Case Collection—these are the 10 company cases in Chap. 3, the people behind the City of Amsterdam, and Executive Director Lincoln Bleveans from Stanford Sustainability, Utilities & Infrastructure for very inspirational dialogues. I also want to thank the Nordic Circular Hotspot for being a strong network on ciruclarity and to their contribution.

I also want to thank my family for their continuous support in climbing the mountain it was to write these books. Special thanks to Søren Krasilnikoff for being my private editor providing the push and the critics needed, and most importantly building the trust in me to fullfill this project. Thanks to all my four children for being the reason and constant inspiration to pursue dramatic change and having the courage to speak up.

References

Haar, G. (2024a). *The great transition to a green and circular economy. Climate nexus and sustainability.* SpringerNature.
Haar, G. (2024b). *Rethink economics and business models for sustainability. Sustainable leadership based on the Nordic Model.* Springer.

Copenhagen, Denmark   Gitte Haar

August 2024

Contents

Introduction

The Nordic Region including Norway, Sweden, Denmark, Iceland, and Finland is a group of countries with societies that have been proven strong, stable, and prosperous and with a common share of social welfare, as well as a high level of social security for not only citizens but also the business environment. Not only the societies in the Nordics are receiving attention these years, the business environment and business conduct here are also different and learnings can be made from this in the transition to a green and circular economy creating a fair and sustainable planet. The traditions of Nordic business conduct are based on a flat governance structure, open and frank discussions, and an informal tone across all levels in the organization, a lack of authority, a high degree of trust, and social interactions across levels in scoiety. All the people here grow up in the same free educational systems in these small countries, and the next boss often is an old schoolmate.

© The Author(s), under exclusive license to Springer Nature
Switzerland AG 2025
G. Haar, *Nordic Case Collection on Sustainability and Transition to a Circular Economy*, Springer Business Cases,
https://doi.org/10.1007/978-3-031-78638-9_1

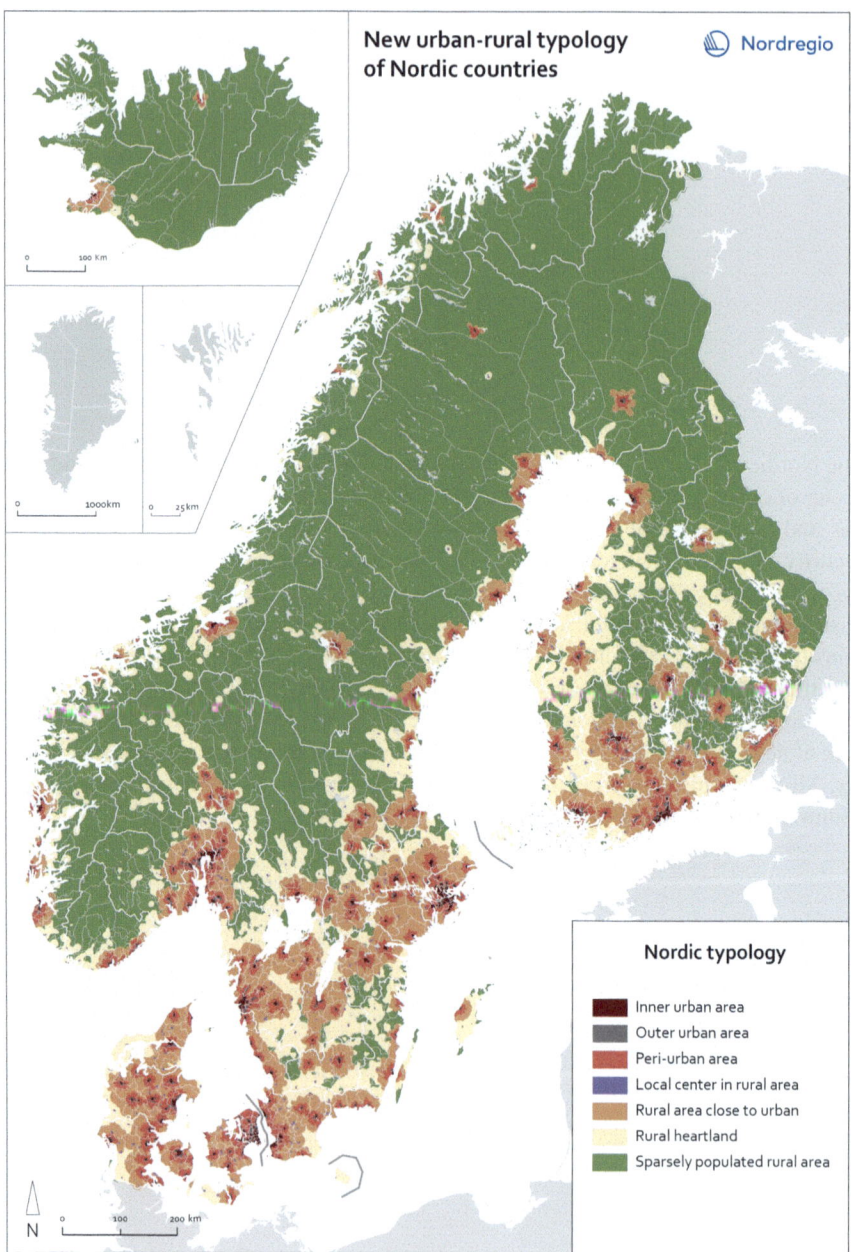

Map of the Nordics. https://www.nordregio.org/maps/new-urban-rural-typology-of-nordic-countries/ (Vasilevskaya and Penje 2023)

The Nordic Region and its characteristics are described in the book *Rethink Economics and Business Models for Sustainability based on the Nordic Model* (Haar, Rethink Economics and Business Models for Sustainability. Sustainable leadership based on the Nordic Model, 2024c).

The Nordic societies with a high degree of equality are based on:
- Trust in the people, the state, the judiciary, and the democratic institutions (governmental and non-governmental)
- High taxation to spread wealth and create equal access to basic needs
- Tax-funded education, healthcare, social welfare & benefits, and infrastructure
- Safe societies with low crime rates
- Economic stability, also in times of international financial crises

The Nordic countries are small countries with small populations. Apart from the dark winters and cold weather, the Nordic people share common values in their definition of a decent and free life, very different from how, for example, Americans define freedom. These shared values make economic equality possible to a larger extent than in many other countries. The economies and the welfare of the Nordics are more stable and extensive than in many economies around the world. The basics of the Nordic society model are illustrated in Fig. 1.1 (Haar, Rethink Economics and Business Models for Sustainability. Sustainable leadership based on the Nordic Model, 2024c).

The people of the Nordics may have developed the resilient societies as they evolved from the rough nature, the shifting seasons and the dark, cold and snowy winters that requires preparing and storage of resources for the winter season with very little available. This is still seen from the large wild mammals, rodents, birds and threes

Fig. 1.1 The basics of the Nordic society model. The model illustrates that trust is the fundament in the Nordics and causes the spread of wealth, with publicly financed education, healthcare, social welfare and infrastructure, high general level of education, and a safe society with low rates of crime

indigenous to the boreal forests of the Nordics. Adaptation is unique to the local climate and this might also have influenced the people and the development of urban societies of today.

The Nordics have some large corporations on the global scene that not only out-perform other corporations in business, but are also in front when implementing sustainability. Read more about the Nordics and the SDGs from a Nordic perspective in another book by the same author: *Rethink Economics and Business Models for Sustainability. Sustainable Leadership based on the Nordic Model.*

Social benefits in the Nordics are strong due to the long tradition of regulating according to human rights as well as strong agreements between corporations and unions in securing employee rights, health, and safety. Social impacts in the value chains of the Nordic consumption especially outside the EU need improvements and an extended social responsibility, especially the large businesses dependancy on manufacturing in Asia that does not always comply with the international human and social rights of the new reporting regulation in the EU (CSRD).[1] The extended responsibilities of the value chains that are implemented with the new Sustainable Due Dilligence Directive in the EU (CS3D[2]) will put legislative responsibilities for social impacts in scope 3 on the large European corporations.

The Nordic business environment and the Nordic societies focus on sustainability and some of the global companies here are nominated and awarded for their sustainability efforts and their disclosures on Environment, Social, and Governance (ESG) as presented in this Case Collection. Many of these companies commit to Science Based Targets Initiative (SBTi[3]), and some of these companies have just now gotten their targets and plans for minimizing climate impacts in scope 3 rejected by SBTi. Greenhouse gas (GHG) emissions and environmental impacts on ecosystems in scope 3 are very significant and for most companies these impacts account for more than 80% of total emissions and impacts. Therefore, it is important that companies start monitoring and regenerating their impacts in scope 3, as required in the new European legislation on reporting (CSRD, SFR[4]), circular economy, and environmental product regulation (ESPR[5]).

The Nordics are large consumers of products and producers of waste globally, and this consumption has huge impacts on the environment and climate change. Here dramatic changes are needed to counter these impacts. These patterns of over-consumption put the companies in the center of driving the changes needed here, as

---

[1] https://finance.ec.europa.eu/capital-markets-union-and-financial-markets/company-reporting-and-auditing/company-reporting/corporate-sustainability-reporting_en

[2] https://commission.europa.eu/business-economy-euro/doing-business-eu/sustainability-due-diligence-responsible-business/corporate-sustainability-due-diligence_en

[3] https://sciencebasedtargets.org/

[4] https://finance.ec.europa.eu/regulation-and-supervision/financial-services-legislation/implementing-and-delegated-acts/sustainable-finance-disclosures-regulation_en

[5] https://commission.europa.eu/energy-climate-change-environment/standards-tools-and-labels/products-labelling-rules-and-requirements/sustainable-products/ecodesign-sustainable-products-regulation_en

they control the supply chain downstream and influence consumption patterns upstream—all in scope 3. So, even if the Nordics stand out in the green transition and sustainability on society and company levels there is still much to be done here in transforming to a circular economy to achieve responsible consumption, resource efficiency, abolition of waste, and becoming independent of virgin raw materials including the retention of critical raw materials for the industries here.

EU Green deal includes a strategy and action plan (CEAP) to transform the European economy to a clean and circular economy, and so does the Nordic Counsil's commitment to climate and environmental action.

1.1   Sustainability in the Nordics

An important stakeholder in the Nordic Region is the Nordic Council. The Nordic Council is the official body for formal inter-parliamentary co-operation. Formed in 1952, it has 87 members from Denmark, Finland, Iceland, Norway, Sweden, the Faroe Islands, Greenland, and Aaland.

The Nordic Council is a regional body uniting the Nordic countries and includes councils of the different profession ministries—such as the Nordic Council of Environmental Ministers. This is to develop and align politics and markets within the Nordic Region, but they hold more of an advisory role than the EU governing bodies.

The Nordic Region has a vision to become **the world's most sustainable and integrated region by 2030**. Cooperation in the Nordic Council of Ministers will serve this purpose and includes (https://www.norden.org/en/declaration/our-vision-2030):

- **A green Nordic region** – Together we will promote a green transition of our society and work for carbon neutrality and a sustainable circular and bio-based economy.
- **A competitive Nordic region**—Together we will promote green growth based on knowledge, innovation, mobility, and digital integration.
- **A socially sustainable Nordic region**—Together we will promote an inclusive, equal, and cohesive region with shared values and strengthened cultural exchange and welfare.

The Nordics are in the transition to meet the Paris Agreement in scope $1 + 2$ at a national and regional level by transforming the energy sector into renewable energy supply. Historically most of the Nordic countries have been provided by hydropower, biomass, and nuclear energy, except Denmark where the energy production historically came from coal, oil, and gas. Incineration of waste and biomass for energy has also been an important energy source in district heating in cities in large parts of Denmark and Sweden. Especially Norway and Denmark have extracted large amounts of fossil fuels (oil and gas) from the North Sea that have been exported out of Norway and created export and self-support of energy in Denmark.

Now the need for resource efficiency and circular economy in the manufacturing industries is more urgent than the transition of the energy industries. There is a need for phasing out waste as an input for energy production and a transition of the district heating system to renewable sources. The GHG emissions from electricity are declining in all countries, but still there is great potential in various industries to:

• Optimize the energy consumption
• Transform to renewable energy
• Electrify industrial processes

for them to meet the requirements set by the EU Green Deal and the vision of the Nordic Counsil, including the new regulation; as Corporate Sustainable Reporting Directive (CSRD) and Sustainable Finance Reporting Directive (SFRD), as well as the new Sustainable Product Regulation (ESPR).

Again, the scope 3 emissions in the Nordics are significant and come from overconsumption of all kinds of goods, but especially clothes, electronics, packaging materials, and building materials are causing tremendous amounts of waste here. This also means that many companies here have more than 80% of their total GHG emissions in scope 3. These scope 3 emissions and impacts are challenging for companies to mitigate, and it requires changing the value chain upstream as well as downstream.

1.2   The Companies in this Nordic Case Collection

This Case Collection contains descriptions of several Nordic companies that have worked strategically with green transition and sustainability for several years. This Case Collection is divided into three and the method of collecting the cases are different as descibed here:

• **Global Nordic corporations** are reviewed based on public available information. The corporates presented in Chap. 2 have marketed themselves on sustainability or have a significant influence on the society and business environment here. Some of the corporations here have been nominated and granted sustainability awards internationally as predecessors in sustainability. Their external sustainability reporting and the web of these companies have been reviewed and an assessment has been made of their sustainability profile and their level of maturity in the green transition. Many of these corporations may have strategic initiatives and work internally on actions, new business models, or products that have not been announced yet and that the public and the author is not aware of. The companies are included in the Climate Nexus desciptions in Chap. 2.
• **Small and medium-sized enterprises** that are in the transition to a circular economy and sustainability in a strategic manner are presented in Chap. 3. They are significant because they have made sustainability and especially circular economy an integrated part of their business for several years. The case compa-

nies here have provided their own take on the benefits and challenges of their work on the green transition and circular economy. The author, has provided each case description with a comment and an assessment of their maturity level in a circular economy. These are included in Chap. 3.

- **Outliners** that have inspired in significant ways in their approach to the transition and their interacting and engagement with stakeholders and the full value chain in a visionary and extraordinary way. The two outliners are not from the Nordics but are important as inspirators and therefore included in this Case Collection. They have provided their case descriptions and the author has made an introduction to the outliner cases in Chap. 4.

References

Haar, G. (2024c). *Rethink economics and business models for sustainability. Sustainable leadership based on the Nordic Model*. Springer.

Vasilevskaya, A., and Penje, O. (2023). New urban-rural typology of Nordic countries. Nordregio.

Climate Nexus and Corporates in the Nordics

The impacts from the Nordics including all three scopes (scope 1 + 2 + 3 according to UN Greenhouse Gas Protocol for Companies[1]) and are reviewed in the framework of the Climate Nexus to describe the transition level, challenges and review selected corporates from relevant sectors. The environmental impacts are the main subject to review here, as the social impacts in the Nordics are highly covered by regulation and the public social welfare and healthcare so significant for this region. The Climate Nexus is illustrated in Fig. 2.1 and explained below. With the Climate Nexus descriptions are also descriptions of the sectors and large corporations involved in the slices of the Climate Nexus.

The Climate Nexus is used to assess some of the Nordic corporates that operate within the industries defining the areas in the Climate Nexus. This chapter includes reviews of selected corporations to give a picture of the state and speed of the transition to a green and circular economy. The cases included in this chapter are based solely on publicly available information such as sustainability reporting, annual reports, and websites.

---

[1] https://ghgprotocol.org/

© The Author(s), under exclusive license to Springer Nature
Switzerland AG 2025
G. Haar, *Nordic Case Collection on Sustainability and Transition to a Circular Economy*, Springer Business Cases,
https://doi.org/10.1007/978-3-031-78638-9_2

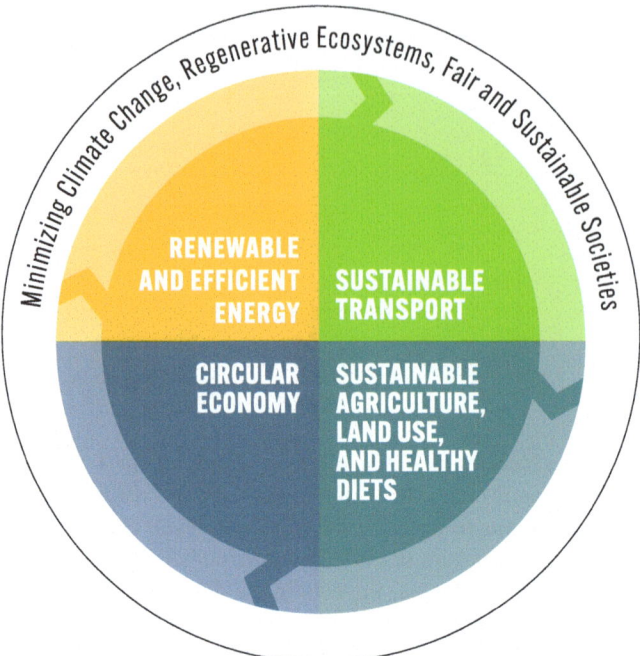

- Political instability from uneven fossil energy source
- Insecure energy supply
- Centralized energy supply
- Particle pollution
- Increasing and unstable fossil energy prices

- Energy inefficient technologies taking over (ICE)
- Particle pollution
- Noise pollution
- Congestion of traffic - time pollution
- Unlivable Cities

- Increasing prices on raw materials
- Resource Scarcity
- Linear business models: Take-make-waste
- Overproduction of building materials, textiles, medicine, etc. never reaching the customer
- Long unstable value chains

- Food waste
- Loss of wild nature
- Loss of biodiversity
- Loss of carbon sources and storage in woods
- Unhealthy diets from industrialized food production
- Over production of meat and inefficient calorie production
- Obesity

Fig. 2.1 Climate Nexus. This Climate Nexus visualizes a holistic and overall approach toward environmental impacts and emphasizes the necessity to understand the complexity of climate impacts. The Climate Nexus is the overall framework of the book: *The Great Transition to a Green and Circular Economy*. Gitte Haar. Springer 2024. Where the details of the Climate Nexus are explained

2.1   Transition to Renewable and Efficient Energy

Transition to renewable and efficient energy mainly minimizes impacts of the Nordics territories and is the scope 1 + 2 of the Nordic countries. The Nordics are well on track with the commitments and plans already committed to by the governments of the five Nordic countries. Some of the Nordic countries have strong traditions in energy production from hydro- and geothermal power technologies and have for many years produced low climate impact energy. Denmark and Norway are oil extracting countries and especially Denmark has historically been powered by fossil energy sources. Independent of the technologies in the energy sector, there is still a need to focus on the energy transition in the Nordics, both for the corporations that for a long time have been producing hydropower, and for the corporations that are still using fossil energy sources. There is still a need for extending and upgrading the power infrastructure technically and digitally, to develop and implement new storage and transformation technologies such as batteries and other Power2X technologies. The region is still seen as a frontrunner in the transition to efficient and renewable energy supply, as well as having the digital infrastructure to efficiently manage energy consumption.

A large part of scope 3 climate impacts come from the supply of imported goods and technologies and here the extraction and processing of materials and products account for the significant part of climate impacts. These emissions represent two types of emissions related to the value chain:

- Climate neutrality from the manufacturing and supply chain (downstream)
- Reuse and recycling of products and materials in the full value chain upstream as well as downstream

The energy savings are huge from reuse and recycling of products and materials compared to harvesting and processing from virgin raw materials. Dependent on material type, recycling saves between 30 and 90% of GHG emissions compared to extraction and wasting virgin materials. This is why the introduction of a circular economy is so important. Read more about this later in this chapter and in a book of the same author: *The Great Transition to a Green and Circular Economy* (Haar, The Great Transition to a Green and Circular Economy. Climate Nexus and Sustainability, 2024a).

Energy production and infrastructure in the Nordics is dominated by state ownership of a few large corporations. This is also the case in other regions and countries, especially in Europe. In the Nordics some of the energy infrastructure is state-financed and most of the energy producers and providers are state-owned. The Norwegian state owns the oil and gas extracted from the North Sea, whereas the Danish state is licensing the extraction. From the 1960s to 2014, it was reported that 42 billion barrels of oil equivalent (BOE) had been extracted from the North Sea since the production began. The extraction is done by Norway, the UK, Holland, and Denmark. Estimates show that so far half of the reserves have been extracted, and the largest part has been exported supporting the economic development the last few decades in these countries. Oil from the North Sea accounts for 2–3% of global crude oil production.

Denmark, as one of the largest oil producers in the EU, decided in 2020 that no new permissions are granted to search for new oil or gas reserves, and as of 2050 all activity of fossil extraction must be stopped. Denmark suggested at the COP28 (2023) held in the UAE that all oil extraction should be stopped in the North Sea by 2031 to be able to meet the Paris Agreement, but this suggestion found no ground, maybe due to the host nation. Norway has no plan to stop exploiting and extracting oil and gas from new reserves in the North Sea, despite a strong call from UN and others. The UK is granting new permission for oil and gas extraction to become energy independent and still sees fossil energy as a part of the solution also looking beyond 2030 or 2040.

The dominating corporations in the energy sector in the Nordics are presented below. This also includes a presentation of gender equality, since also the Nordics are still struggling with reaching gender equality of a minimum 40% of each gender. Abbreviations used are Female/Male (F/M), Executive Management Team (EMT), and Board of Directors (BoD). Percentages of the least represented gender are stated in (brackets).

2.1.1 Equinor

Equinor
Norwegian oil and gas company from the merger of Hydro (Oil and Gas) and Statoil. They operate globally and the Norwegian state owns the majority of Equinor.
https://www.equinor.com/.
Revenue (2023): 96.5 billion euros. Number of employees: 23,000

Product and business model:
Equinor with its subsidiaries develops oil, gas, wind, and solar energy in more than 30 countries worldwide. A company in the conglomeration is a large waste handling company.

Sustainability:
Vision: "Energy for people, Progress for society, and Searching for better." It is one of the few energy companies in the Nordics that do not have a clear climate or sustainability stake embedded in its vision. Equinor has detailed communication on the transition to **carbon zero by 2050** and 40% reduction in carbon intensity by 2035, and **scope 1 + 2 reductions of 50% by 2030**, including a large part by carbon transport and storage (CCS) based on technologies and installations not yet in place. They communicate minimizing impacts on nature, especially offshore, and zero-harm. They see their license to operate as safeguarding our people, protecting our assets, and a just transition. The focus in Equinor is the energy transition still with a focus on nature and zero-harm. Equinor is very much influenced by the fact that Norway has not decided to stop exploiting new fields of oil and gas. Therefore, Equinor is an energy company with low ambition in the transition plan and their business will still for many years be dependent on extracting, producing, and selling fossil-based energy.
https://www.equinor.com/sustainability

Gender equality in management (F/M):
EMT – 4/9 (31%); BoD – 4/7 (36%).

2.1.2   Ørsted

Ørsted
Danish listed corporation with 50.1% ownership of the Danish state. Originally fully state-owned when being DONG established by the merger of large energy providers in Zealand, Denmark.
https://orsted.com/
Revenue (2023): 10.6 billion euros. Number of employees: 7700

Product and business model:
Operator of energy and energy infrastructure shifting to operating of renewable energy (mainly windmills), globally. From historic activities Ørsted still operates non-renewable energy installations.

Sustainability:
Vision: Building more than ever before. Ørsted is now struggling financially due to development projects in renewable energy that has not proceeded as fast as planned or has the short-term returns that support the business. Unfortunately, they have not been able to create a transition roadmap that is valid and supported by stakeholder, even with the Danish state is the main shareholder.

Ørsted report their ESG ambitions and impacts as an integrated part of their annual report. The transition plan targets **net-zero in scope 1 + 2 + 3 by 2040**. Validated by SBTi. Net-positive biodiversity impact on all new renewable projects commissioned no later than 2030. Zero landfill of wind blades and solar panels. Gender equality target: 40:60 (women:men) in total workforce by 2030. All communicated in the annual report. Ørsted received sustainability nominations and awards for their annual report in 2023. In 2023, Ørsted was awarded the highest possible CDP rating for the fifth consecutive year and recognized as a global leader on climate action and for the seventh consecutive year Ørsted has been recognized as one of the world's 100 most sustainable companies in the "Corporate Knights Global 100 ranking. "For 2024, we rank no. 17 across all industries globally and no. 1 among energy developers" (Ørsted, 2024). This is a little hollow at times where they are writing off and closing the development of future technologies.
https://orsted.com/en/who-we-are/sustainability

Gender equality in management (F/M):
EMT − 1/4 (20%); BoD − 5/5 (50%).

2.1.3   Vestas

Vestas
Danish listed company with equity fonds as main shareholders. Emerged from Danish universities and research and development of wind turbines heavily supported by the Danish state since the 1970s.
https://www.vestas.com/en
Revenue (2023): 15.4 billion euros. Number of employees: 29,000

Product and business model:
VESTAS are designing, manufacturing, installing, developing, and servicing wind energy and hybrid projects all over the world—mainly onshore.

Sustainability:
Vision to become the global leader in sustainable energy solutions, and sustainability in everything they do with four corporate values: Simplicity, Collaboration, Accountability, and Passion.
Extensive communication on environment and sustainability online. Four key goals: i) carbon neutrality by 2030 with the detailed targets: **climate neutral in scope 1 + 2, and scope 3 target of 45% per MWh generated by 2030. No disclosures on full scope 3 neutrality**, ii) zero-waste from turbines by 2040 (unambitious in respect of the need for critical raw materials), iii) becoming the safest, most inclusive, and socially responsible company in the energy sector, and iv) leading the transition toward a world powered by sustainable energy. Approved by SBTi. The goals are not fully supported by sub-targets as climate neutrality is not a full scope 1 + 2 + 3 transition and this is contrasting the target of leading the transition. https://www.vestas.com/en/media/blog/sustainability.

Gender equality in management (F/M):
EMT 0/2 (0%); BoD − 5/5 (50%).

2.1.4   Fortum OYJ

Fortum OYJ
A Finnish listed energy company with the state as a majority owner of 51.26% of the shares. https://www.fortum.com/.
Revenue (2023): 6.7 billion euros. Number of employees: 5000

Product and business model:
Fortum focuses on the Nordics, the Baltics, Poland, and Northwest Russia. Following the acquisition of the Russian energy company TGC 10 in 2008, Western Siberia has become an important business area for Fortum. Fortum produces energy mainly from hydropower and nuclear power. Fortum is investing in wind power and combined heat and power (CHP); they also produce from solar panels to ensure clean energy production. They also provide power trading and energy supply. The goal is to deliver reliable supply of electricity and district heat to private and business customers.

Sustainability:
Purpose: "Our purpose is to power a world where people, businesses and nature thrive together. We help societies to reach carbon neutrality and our customers to grow and decarbonize their processes in a reliable and profitable way, in balance with nature." Fortum communicates their sustainability report and targets are on secure supply, decarbonization, safety, and gender equality. Their strategy and targets are built on SBTi 1.5 °C and **carbon neutrality by 2030 in all 3 scopes and coal exit in their own operations by the end of 2027**, biodiversity, and safety targets. They have started the biodiversity footprint assessment. The sustainability report from Fortum is very extensive and descriptive and links to SDGs. They also have a detailed description on circular economy and acknowledge the transition and their stake in waste management, resource efficiency, resource recovery, and recyclability.
https://www.fortum.com/sustainability

Gender equality in management (F/M):
EMT − 5/6 (45%); BoD − 3/6 (33%).

2.1.5 Statkraft

Statkraft
Norwegian energy provider fully owned by the Norwegian state.
https://www.statkraft.com/
Revenue (2023): 10.5 billion euros. Number of employees: 6000+

Product and business model:
Statkraft is a leading company in hydropower internationally and Europe's largest generator of renewable energy. The Group produces hydropower, wind power, solar power, and gas-fired power and supplies district heating. Third largest energy producer in the Nordics.

Sustainability:
Vision: Renew the way the world is powered. The sustainability approach at Statkraft is anchored around climate, biodiversity, human rights, and circular economy. They commit to Paris Agreement 1.5°C goals for the energy sector with detailed goals in all three scopes, commit to the Global Biodiversity Framework (GBF) from COP15, and have targets on monitoring. They have commitments to Human Rights and ILO (International Labour Organisations) on specific targets and want to make a positive impact on circular economy; they commit to start mapping, investigating, and monitoring impact and pilot projects but with no measurable goals on circularity. The company is targeting carbon neutrality for **its scope 1 and scope 2 emissions by 2040**. Statkraft also aims to reduce emissions from its supply chain and will encourage suppliers to contribute to this effort. The sustainability report is not as detailed and thorough as other large corporations, but the topics touched upon are relevant. It seems as if they are in the beginning of onboarding sustainability and the climate change transition plan, and they have the right focus. /https://www.statkraft.com/globalassets/z_unlisted-documents/sustainability-strategy/sustainability-strategy-2023.pdf.

Gender equality in management (F/M):
EMT – 4/4 (50%); BoD – 4/5 (44%) Female CEO and chair.

2.1.6 Vattenfall

Vattenfall
The group is 100% owned by the Swedish state.
https://group.vattenfall.com/
Revenue (2023): 25.5 billion euros. Number of employees: 21,000

Product and business model:
One of Europe's largest producers and retailers of electricity and heat with main markets in Sweden, Germany, the Netherlands, Denmark, and the UK. Vattenfall produces energy from wind, nuclear, biomass, hydro, solar, gas, coal, and district heating (often waste incineration). In the Nordics (Sweden and Denmark) Vattenfall provides energy produced from wind, nuclear, and district heating. Coal and gas production is in Germany and Holland.

Sustainability:
Mission: "Helping society break free from fossil fuels." Also phrasing a just transition and **with a transition plan to carbon neutrality in 2040**. Their net-zero targets are approved by SBTi. They provide a very extensive and detailed sustainability report with strategic targets and fossil free in the center of the strategy. The report is a roll-out of their full strategy and investment schemes. Sustainability has a very strong anchor in climate neutrality and communicates in two frameworks: ESG and SDGs. They also disclose the result of their materiality assessment and have chosen climate change, biodiversity and ecosystems, circular economy, health and safety, workers in the value chain, and right of indigenous people. The part on circular economy is less extensive, and they disclose historic waste data but no target metrics, just addressing the importance of engaging with supplier. This sustainability report and the way it is organized are impressive and represent the work, documentation, and strategic targets necessary to drive genuine sustainability. The overview is a little scarce but still the sustainability work here is showcasing an example to follow.
https://group.vattenfall.com/investors/financial-reports-and-presentations/
annual-and-sustainability-report-2023/

Gender equality in management (F/M):
EMT – 5/5 (50%); BoD – 3/11 (21%). Female CEO.

2.1.7   Landsvirkjun

Landsvirkjun
National and state-owned by the Icelandic state.
https://www.landsvirkjun.com/
Revenue (2023): 548 million euros. Number of employees: 340

Product and business model:
Landsvirkjun is by far the largest energy company in Iceland, providing approximately 75% of all the electricity produced in Iceland (12.6 GWh annually of total 16.8 GWh), Landsvirkjun produce energy from hydropower, geothermal, and wind.

Sustainability:
Vision: A sustainable world, powered by renewable energy. Their goal is to **become climate neutral in 2025** which is an early ambition but applicable due to the high level of renewable energy production today. The vision of Landsvirkjun also focuses around "respecting nature and responsible utilization of resources" and publication of their environmental policy. Landsvirkjun publishes a Climate Action Plan with targets and initiatives and an emphasis on "We Must All Do Better." Their sustainability reporting is immature but with a focus on the most important for an energy company: climate change. Sustainability is anchored around climate neutrality, equality, grants, and community engagement. Circular economy is not an outspoken part of their climate plan, but they still showcase some very interesting cases and investments related to circular economy, but no CE targets in scope 3 or supply chain. Landsvirkjun has been ranked high by the *Financial Times* and CDP with their ambitious climate action plan. Focus on nature is strong in their communication which is natural in a society and landscape that is so wild and dependent on adaptation to nature. Generally, Iceland and its people living on a group of volcanic islands defining not only the Landsvirkjun but also the people here. The national Icelandic energy company leaves an impression of having had climate change and nature at the heart/core of the business conduct for many years, and thereby the communication on this is less expressive as seen by some of the larger energy corporations in Norway, Sweden, and Denmark.
https://www.landsvirkjun.com/climate-action/climate-action-plan

Gender equality in management (F/M):
EMT – 4/5 (44%); BoD – 2/3 (40%).

The energy sector in the Nordics is mainly state-owned which strongly influence the high ambitions. For historic reasons, the energy transition of the Nordics is far on track to an efficient, clean, and renewable energy according to the yellow part of the Climate Nexus (Fig. 2.2). Nevertheless, the Danish transition to 2030 target on 70% reduction is in danger of not being met as the transition to renewbale energy has almost stopped here. The main reason is political and bureaucratical resistance for private and companies to install. Recently the Danish state just experienced no bidders on a tender on a new ofshore wind turbin park and this will even further challenge the 2030 target. The rest for the Nordic countries are on track here.

All Nordic countries hold climate transition plans even if the national goals are not aligned. The energy producers as a minimum meet the national and EU targets on climate neutrality, and in Iceland, Sweden, and Finland the transition of the energy sector is on track or ahead. As Denmark and Norway are oil and gas extracting countries, the transition is moving slower here. Here the transition plan for Finland is included in Fig. 2.3, because it stands out as the most ambitious plan in the Nordics (Finland, 2024).

The Finnish transition plan has an ambition to become carbon sink by capturing 25 MtCO$_{2e}$ carbon, from regenerating land use into a large sink from being a small sink today. This will hopefully also be implemented in respect of internationally commited ecosystems and biodiversity targets.

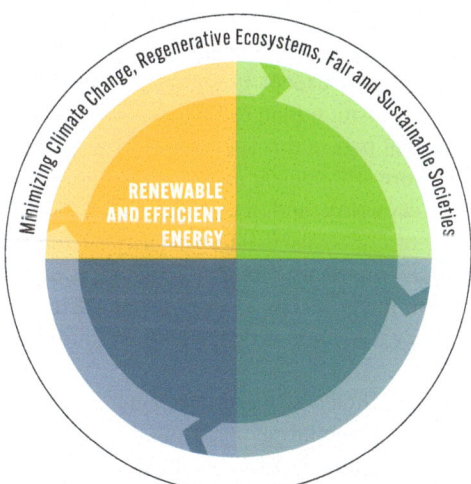

Democratic and free access to energy for all resulting in political stability, spread of democracy, and minimizing poverty.

Predictable prices on energy supply as an important resource for consumers and business.

Eliminating pollution from combustion engines That causes sever health care problems, from particles and noise.

Fig. 2.2 Climate Nexus—Renewable and efficient energy. The Climate Nexus is a concept described in the book: The Great Transition to a Green and Circular Economy. Gitte Haar. 2024 and visualizes a holistic approach to implementing sustainability

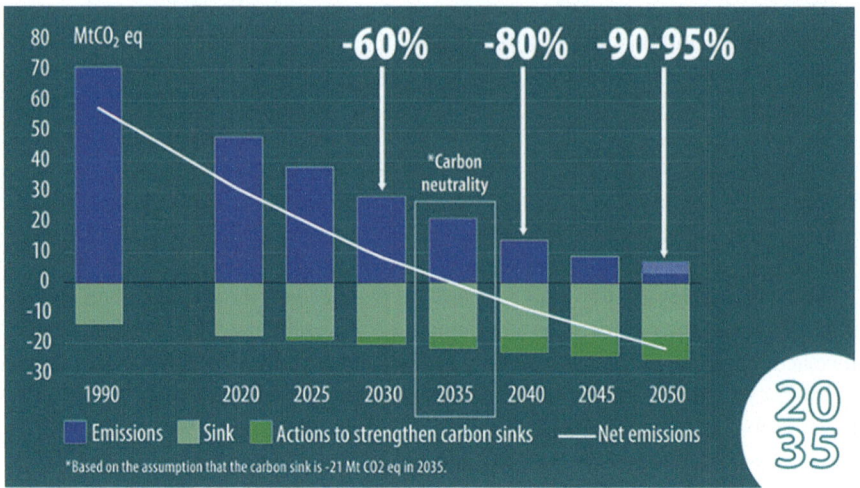

Fig. 2.3   Finland transition plan. Finland is probably the most ambitious country in the Nordics and has approved a reform for Finland including very ambitious targets in sustainability including this target to become climate neutral by 2035

Generally, the Swedish energy corporations seem to be the most ambitious on the energy transition with the most extensive plans and targets in all three scopes.

All the energy corporations, except Equinor, presented above hold targets to become fossil free and a transition plan to become climate neutral in scope 1 + 2, which is the most important part in this sector. Some also have targets for scope 3, but the targets in scope 3 are still unclear and needs updating to comply with future legislation on circular economy and the requirements to a transition plan in CSRD. The targets are rarely focused on circularity of the installations and the retaining of the critical raw materials. Circular economy is important to support the energy transition because many resources are embedded in the renewable energy installations and the energy transition is dependent on future installations of renewable energy installations.

The potential of energy efficiency and minimizing energy consumption in various industries as well as the potential savings from the recycling of materials is still large and connected to the implementation of a circular economy.

Denmark is generally falling behind in implementing renewable energy and has not as Sweden and Germany updated its goal on climate neutrality to 2045. Finland has the goal to become climate neutral by 2035. Norway is aiming for carbon neutrality by 2030 (scope 1 + 2). Iceland aims for climate neutrality by 2040. Denmark still sticks to the original EU goal on climate neutrality in 2050. These ambitions are reflected in the different energy companies, since the energy sector in the Nordics is state-owned and closely monitored.

Aligned ambitions and transition plans for climate mitigation and adaptation in the Nordics as a region would be a strong political signature, especially because the politicians in the Nordics see the countries and the region as frontrunner in the green and circular transition.

2.2 Transition to Sustainable Land Use and Healthy Diets

Transition to sustainable land use and healthy diets is important because GHG emissions from land use are a large part of the GHG emissions in the Nordic geography (scope 1 + 2). Significant impacts on climate, land and water ecosystems, and biodiversity are created indirectly from imports of feed, meat, and fertilizers that all have a strong negative GHG impact and a negative impact on biodiversity and ecosystems outside the Nordics (scope 3). Land use, agriculture, and forestry also have significant impacts on nature and the ecosystems in the Nordics, such as biodiversity, ocean and freshwater ecosystems, and the extraction of clean drinking water here. It is important to understand that land use and agriculture is also impacting ocean and freshwater ecosystems and that natural ecosystems on land and in water are strongly linked and connected. This is why this part also includes a review of the marine and freshwater ecosystems and challenges here.

Especially in Denmark highly industrialized and intensive agriculture dominates more than 60% of the land of Denmark. Thus, Denmark and parts of southern Norway and Sweden are suffering from a low degree of wild nature and low biodiversity. Agriculture and land use is an important emitter of GHG in scope 1 + 2 + 3 and causing huge environmental impacts on the coastal areas, oceans, and freshwater ecosystems. Finland and Sweden hold a strong forestry industry, and especially the ecosystems, nature, and biodiversity of Finland are suffering from the intensive production of wood. Denmark and Finland are both poor in biodiversity on land and do not meet the EU requirements for wild nature and biodiversity, whereas Norway, Sweden, and Iceland hold huge areas of wild nature.

Most of the Norwegians and the Swedes live in the southern part of the countries where land is suitable for agriculture. In the Nordics, urbanization is strong and between 80 and 93% of the Nordic people live in cities, except for the Finns where 70% of the population lives in cities. The Nordic societies are representing the urbanization that is expected globally with more than two-thirds of the population to be living in cities in the future. Urban living is often an efficient way of living if we can provide food, energy, goods, and transportation in a sustainable manner.

Denmark including Greenland, Norway, and Iceland has a strong fishing tradition and has large fleets of fishing vessels and industrialized fishing. This has resulted in overfishing not only from the local fishermen but also from other parts of the EU and Great Britain. Now EU commitments and national legislation, as well as new agreements after Brexit, hopefully will change the pressure on the North Sea, the Baltic Sea, and the inland waters of Scandinavia. Norway also has a large industry of fishing and fishery farming that also impacts the North Sea due to pollution from feed and medicine (antibiotics) to the fish in the aquacultures. Thereby the oceans and freshwater systems in the Nordics are challenged by an intensive fishing industry but also an intensive agricultural production and the keeping of livestock in Denmark leach huge amounts of nutrients (nitrogen and phosphorus) to the oceans and freshwater systems causing oxygen depletion and the death of the fish stocks here. This year (2024) the inner waters of Scandinavia intensively suffered from oxygen depletion, grease, and decline of ocean and freshwater ecosystems including the fish populations that have provided food and wealth in the Nordics for centuries. The water ecosystems are now in such a poor state that fishery has been put on hold for years for the biodiversity of fish and marine ecosystems to bloom again, and to reintroduce gently and more passive fishing tools, as used historically.

The cold waters of the North Sea have historically been a significant source of food for the people here and this has now come to an end. Around Greenland in the arctic waters pelagic fishing is not as threatened as further south in the North Sea and the Baltic Sea, but all in all dramatic changes are needed to restore one of the world's most important and significant ecosystems and food systems, not only for humans but also for the mammals living here on land and in the oceans, such as whales, polar bears, and other smaller mammals that are dependent on the oceans as their primary habitat.

The rapid melting of ice around the North Pole and Greenland is causing a rapid change of the ecosystems, affecting not only the region itself but also the global ecosystems and food systems. These dramatic challenges from overproduction of food and climate change are now receiving the necessary attention from politicians here and within Europe, but not with the ambitions and speed necessary to restore and regenerate the ecosystems.

Generally, the transition to a sustainable agriculture is on the agenda in the Nordics. The intensive production of meat from poultry, cows, and especially pigs is a big issue especially in Denmark but also in the southern part of Sweden. In Denmark, meat production is very industrialized and intensive, and the manure together with the intensive use of fertilizers is harming nature, biodiversity, and water ecosystems. The EU has set targets on wild and protected nature, biodiversity, share of organic farming, and phasing out of fertilizers, pesticides, and microbial antibiotics. The discussion in Denmark is very much around a carbon tax on agriculture and an agreement has been made to meet the committed targets to climate neutrality, wild nature, marine and freshwater ecosystems, and biodiversity.

Denmark has a political agreement and targets to:

• Raise 250.000 ha new forest to increase carbon capture and biodiversity
• Take 140.000 ha of wet-land soils out of agricultural production to minimize methane emissions
• Protect and create a cleaner freshwater system to regenerate clean drinking water

No action plan or specific targets or actions have been agreed upon or is in implementation, yet (2024). As the very intensive primary producer of food, especially the Danish approach to regenerate wild nature is important in the southern part of the Nordics and around the Baltic Sea.

The Nordics hold some large food processing companies and selected corporations are reviewed here with links to their published sustainability impacts and targets (April 2024). Gender equality in management is noted after each company description with (Female/Male), Executive Management Team (EMT), and Board of Directors (BoD).

2.2.1   Arla

Arla
A large dairy corporation owned by farmers in Denmark, Sweden, the Great Britain, Germany, Holland, Belgium, and Luxembourg
www.arla.com
Revenue (2023): 13.7 billion euros. Number of employees: 20,900

Product and business model:
Arla collects milk from their owners and produces milk and cheese products as well as ingredients from whey.

Sustainability:
Vision: Creating the future of dairy to bring health and inspiration to the world, naturally. Arla has released their roadmap on the climate ambitions with **targets of 63% reductions in scope 1 + 2 and 30% in scope 3 by 2030, and climate neutral in 2050**, not particularly strongly addressing the farm challenges from cows as a significant emitter of methane (CH_4), nor disclosing a transition or action plan to meet the targets. Approved by SBTi according to Arla's own web. The roadmap is detailed on the 2030 targets, but the large part of the emissions on farm, cows, and land use only are addressed by some highlights that must be unfolded intensively to meet the 2050 target of climate neutrality. Carbon sequencing is part of the actions highlighted and Arla states the purpose is to balance GHG emissions with carbon removals to reach carbon net zero. This is a huge task, and the actions needed here are extensive as the methane emission from milk and beef production is massive. https://www.arla.com/sustainability/arlas-climate-ambition/

Gender equality in management (F/M):
EMT − 1/7 (13%); BoD − 4/15 (21%).

2.2.2   Danish Crown

Danish Crown
A large slaughterhouse group owned by 5600 Danish farmers and thereby representing a large part of the value chain in the meat industry.
https://www.danishcrown.com/en-uk/.
Revenue (2023): 5.5 billion euros. Number of employees: 23,000

Product and business model:
DC is the world's largest exporter of pork meat and Europe's largest producer of pork. The Danish Crown Group is also Europe's largest meat processing company, and the Group is a significant player in the European beef market.

Sustainability:
Vision: We take the lead. Stating its **ambition is to become climate neutral by 2050** engaging with the pig farmers: their sustainability communication online is very much about storytelling based on issues addressed within the framework of farm to fork—a sustainable food production, a diverse and safe workplace, sustainability from slaughterhouse to consumer, animal welfare, and biodiversity. Under these topics DC tells stories on relevant subtopics but with no targets, concrete actions, or transition plan. It is barely good intentions. This reporting does not comply with the requirements of CSRD and does not bring comfort to stakeholders that they have a plan to transform. For many years, their strategy was called a four-wheel strategy addressing their main stakeholder farmers driving their beloved tractor. The climate target is net zero on meat production by 2050, and setting science-based targets is a concrete step toward this target. DC commits to reduce Scope 1 + 2 GHG emissions by 42% by 2030 from a 2020 base year. DC commits to reduce Scope 3 GHG emissions with 20% per kg of output produced by 2030 from a 2020 base year. SBTi approved. Generally, there is a lot of general information on sustainability topics on the website but without a transition plan and without giving an overview of actions and detailed targets on other topics than climate neutrality. It almost seems as if DC has published their own educational material in understanding sustainability topics which might be relevant for their stakeholders, but the commitments and ESG reporting are very poor or almost not existing. It does not bring comfort to the public that they are able to implement the roadmap to meet the targets. Danish Crown was sued by a Danish Climate Act movement for greenwashing in marketing climate-friendly pork. The lawsuit came out even and has been put to the High Court for a final decision. DC just had a new CEO and hopefully he will put sustainability in a more strategic position rather than a communication position as it is now. https://www.danishcrown.com/global/sustainability/

Gender equality in management (F/M):
EMT – 0/2 (0%), 1/3 (25%) including staff management, 1/8 (11%) including management in subsidiaries; BoD – 1/10 (9%).

2.2.3 AAK AB (AArhusKarlshamn)

AAK AB (AArhusKarlshamn)
AAK is a fusion of a Danish and Swedish public listed company in Sweden.
https://www.aak.com/
Revenue (2023): 3.9 billion euros. Number of employees: 4000

Product and business model:
AAK AB is a multinational producer of vegetable oils. The products are used in food, chemistry, pharmaceuticals, cosmetics, and feed. The headquarters are located in Malmö, and they have production facilities all over the world. In Sweden, they have a production facility in Karlshamn. In Denmark, they have a production facility in Aarhus.

Sustainability:
Vision: Making Better Happen. Sustainability is anchored around the food system and beyond, pursued together with others, emission reduction, deforestation-free supply chain, people, and going forward. Targets are not communicated at the top of the web where sustainability is, but are found deep in pdf sustainability reports. The targets are reduction of **scope1 + 2 emission with 50%, scope 3 non-FLAG emissions with 46%, and scope 3 FLAG emission by 33% by 2030** (Forestry, Land, Agriculture Guidance). **No targets beyond 2030 and no target on climate neutrality and no roadmap to meet the Paris Agreement.** Supported by a People roadmap toward 2030. Approved by SBTi. Like Danish Crown the communication is very much about storytelling, and not fact- and data-based in the initial communication. It still lacks integration into the strategy of the company and still looks mainly like communication efforts. No alignment with EGS (CSRD) framwork.
https://www.aak.com/sustainability/

Gender equality in management (F/M):
EMT – 1/7 (13%); BoD – 2/5 (29%).

2.2.4 Scandi Standard AB

Scandi Standard AB.
Privately owned corporation listed in Sweden.
https://www.scandistandard.com/
Revenue (2023): 1.2 billion euros. Number of employees: 3000+

Product and business model:
Producing various chicken products in the Nordic Region and Ireland, and production and sale of eggs in Norway. The chicken brands under Scandi are Danpo, Den Stolte Hane, Kronfågel, Manor Farm, and Naapurin Maalaiskana, which can be found in over 40 countries.

Sustainability:
Vision: "Our aim is to assume an industry-leading role in animal welfare, as well as in environmental and social responsibility." The sustainability report is anchored around people, chicken, planet and products. They commit to Global Compact and Agenda 2030—the SDG#2, 3, 12, 8, 13. They link to various policies. The link to the sustainability report did not work. The communications on the web are valid and deep. In the environmental policy are some climate targets: reduce **scope 1 + 2 GHG emissions 50% by 2030**, reduce **scope 3 GHG emissions from purchased goods and services, fuel- and energy-related activities, and upstream transportation and distribution with 50% by 2030,** and the target boundary includes biogenic land-related emissions and removals from bioenergy feedstocks. **No target on climate neutrality**, no transition plan to Paris Agreement, and no clear roadmap on the chicken production targets. Signed SBTi. As other corporates in the food sector, communication on sustainability is more about storytelling than fact-based and adapted to the new CSRD regulation. They state to work on animal welfare but not clear targets or actions could be found.
https://www.scandistandard.com/sustainability/

Gender equality in management (F/M):
EMT – 1/11 (8%); BoD – 2/6 (25%).

2.2.5   Lantmännen

Lantmännen
An agricultural cooperative owned by 18.000 Swedish farmers.
https://www.lantmannen.com/
Revenue (2023): 1.89 billion euros. Number of employees: 12,000

Product and business model:
Lantmännen operates within agriculture, machinery, bioenergy, and food products.
Lantmännen is one of Northern Europe's leading actors and is active in more than 20
countries. The base is grain products and Landmännen refines the field's resources into a
viable agriculture. Some of the well-known brands are AXA, Schulstad, Hatting, AMO, FINN
CRISP, and Kornkammeret.

Sustainability:
The vision: "Responsibility from field to fork." They state that the company is founded on
knowledge and values that have existed for generations with our owners and through research,
development, and activities throughout the value chain; they take responsibility from farm to
fork. Lantmännen publishes a reasonable but overall sustainability disclosure, with **climate
targets of neutrality in the value chain by 2050 and 50% from scope 1 + 2 by 2030,** and
targets for purchased transport, also addressing climate and ecosystems, sustainable products
and renewable raw materials, and responsible employer and business partner.
https://www.lantmannen.com/sustainable-development/.

Gender equality in management (F/M):
EMT – 3/8 (27%); BoD – 4/10 (29%)

2.2.6   Orkla Foods

Orkla Foods
Privately owned corporation listed on the Oslo Stock Exchange, Norway.
(https://orklafoods.com/)
Revenue (2023): 5.76 billion euros. Number of employees: 19,700

Product and business model:
A large Nordic corporation with many consumer food brands within several categories, such
as food, snacks, confectionery, biscuits, health, personal care, textiles, detergents, and painting
equipment. Their markets are in the Nordics, Baltics, and Central Europe, and in several of
these countries, they hold a leading market position in categories like frozen pizza, ketchup,
soups, sauces, bread toppings, and ready-to-eat meals. The products are primarily sold through
the grocery channel, but they also hold strong positions in the Out of Home, convenience
store, and petrol station sectors. Thereby, Orkla Foods is a significant player in the value chain
of food products.

Sustainability:
The sustainability vision is: "We create positive change by enabling a responsible transition
towards net zero and sustainable production and consumption." They publish short
sustainability communication in the ESG framework with a clear transition plan **with
commitment to net zero in 2045 and with scope 1 + 2 reduction target of 70% by 2030.**
Linking to sustainability reporting and policies.
https://www.orkla.com/sustainability/

Gender equality in management (F/M):
EMT – 2/10 (17%); BoD – 3/3 (50%).

2.2.7   DLG

DLG
A cooperative owned by 26,000 Danish farmers
https://www.dlg.dk/
Revenue (2023): 11.6 billion euros. Number of employees: 6300

Product and business model:
DLG is a leading company in primary production for food and feed, energy (biogas), and housing with business around agricultural grain production for human and animal food and energy produced for manure from farms.

Sustainability:
Vision: Progress and sustainability in agriculture and the food sector. Their climate ambition is that group activities will be **climate neutral by 2050 scope 1 + 2 + 3 including farm level**. Neutrality in 2050 for a corporation based on grain production seems unambitious. The communication is poor with six selected SDGs – SDG#2, 5,8, 12, 13, and 17. DLS supports Global Compact, and the communication still needs to be updated to the new legislation (CSRD). No mention on biodiversity, wild nature, and land use was found by scrolling the web, except for the deforestation-free soya certification. DLG is involved in new sustainability projects, such as BioCirc a new energy cluster based on renewable energy and biogas in Denmark, production of vegetable-based proteins for feed, and others. Initiatives for sustainable solutions are in the scope of their business.
https://www.dlg.dk/Bæredygtighed/

Gender equality in management (F/M):
EMT – 2/4 (33%); BoD – 3/10 (23%).

2.2.8   Cloetta AB

Cloetta AB
Privately owned and listed in Sweden
https://www.cloetta.com/en/
Revenue (2023): 704 million euros. Number of employees: 2600

Product and business model:
A leading confectionary company in Northern Europe. The products are sold in more than 60 countries worldwide with Sweden, Finland, Denmark, Norway, the Netherlands, Germany, and the UK as the main markets. The brands are Läkerol, Cloetta, CandyKing, Jenkki, Kexchoklad, Malaco, Sportlife, and Red Band. Thereby, Cloetta is a significant procurer of sugar, cocoa, and palm oil which are important crops, globally.

Sustainability:
Overall vision for sustainability is "to create long-term value and impact. Sustainable value is about growing the company and at the same time ensuring that the people and environments that are affected by Cloetta's operations or products are positively impacted." They disclose sustainability communication on top of the web with a good sustainability disclosure and targets on biodiversity, social impacts for cocoa farmers, and a **46% absolute GHG emissions reduction by 2030**, but **no climate neutral target and no transition plan disclosed**. The planetary targets also include targets on packaging and responsible sourcing of raw materials (cocoa and palm oil). They have targets for customers (for you) such as sugar-free, less sugar and options with functional ingredients, more vegan options, and supporting dental health with our xylitol products. Their people strategy is based on no work accidents, purpose-driven community engagement by 2025, and partnerships to improve living conditions in our supply chain by 2025. As a significant procurer their planetary / environmental targets and efforts could be strengthened to show that they drive transition rather than rely only on existing certificates.
https://www.cloetta.com/en/sustainability/

Gender equality in management (F/M):
EMT – 2/8 (20%); BoD – 4/4 (50%). Female CEO.

Forestry.
The largest forestry corporations are Stora Enso, UPM-Kymmene, and Metsä Board, all Finnish, and they represent a significant part of the paper, cardboard, packaging, and furniture industry in Finland dependent on industrialized forestry. They often own or rent forest where they source their wooden inputs from, as well as they produce energy from the wood residues. Only one of these large operators is reviewed here:

2.2.9   StoreEnso

StoreEnso
StoraEnso Oyj is privately owned and is listed on the Nasdaq in Helsinki and Nasdaq in Stockholm.
https://www.storaenso.com/
Revenue (2023): 9.4 billion euros. Number of employees: 22,100

Product and business model:
StoraEnso is a Finnish/Swedish corporation and part of the global bioeconomy, as a leading provider of renewable products in packaging, biomaterials, and wooden construction, and provides packaging, pulp, pellets and paper, wood products, and bio-based innovations.

Sustainability:
Vision: Lead the way towards a sustainable future by providing innovative solutions that contribute to a circular bioeconomy. Their integrated approach to sustainability is to focus on three areas where they have the biggest impacts and opportunities: climate, biodiversity, and circularity. **Climate target is to reduce absolute CO_2 emissions by 50% by 2030** from the 2019 baseline **in all 3 scopes. They have a target on climate neutrality by 2040 by joining the Climate Pledge initiative.** They do not disclose their transition plan. For circularity, the target is to achieve 100% technically (non-biologically) recyclable products by 2030, and to achieve a net positive impact on biodiversity through active biodiversity management in forests and plantations. Their regenerative ambitions are anchored in their foundation, and they communicate strongly and detailed on efforts and intentions but not as clearly and top-down on targets and commitments to wild nature and biodiversity, and no transition plan is disclosed here. But they communicate about the Montreal Biodiversity Framework. They also use the SDGs as a communication framework. They seem fair on track in understanding and communicating the relevant issues and still need to disclose more target metrics, sub-targets, and transition plans.
https://www.storaenso.com/en/sustainability

Gender equality in management (F/M):
EMT – 3/7 (30%); BoD – 4/4 (50%)

Fishery.
Norway produces two-thirds of the global salmon production and the production methods are very industrialized and intensive and are often criticized for endangering the ocean ecosystems of the North Sea with the overspill of feed and medicine to salmon that affects the natural file of fish and case eutrophication of the oceans. One of the large producers is reviewed here:

2.2.10 Mowi

Mowi
Mowi is privately owned and listed in Norway
https://mowi.com/
Revenue (2023): 5.5 billion euros. Number of employees: 11,500

Product and business model:
Mowi is the largest producer of salmon in Norway and globally they cover approx. 15% of the salmon market, thereby being a very important player in protecting the oceans in the Nordics and other oceans where they operate, and proving food and shareholder value is their top communication. Their business areas are feed, farming, and sales and marketing.

Sustainability:
Vision: Leading the Blue Revolution. Mowi claims to be the most sustainable protein producer and wild fish is a low climate impact protein source, whereas farmed fish has a different environmental impact on climate, ecosystems, and biodiversity. Unfolding the vision with: "Our goal is to provide a growing world population with delicious, healthy and nutritious food from the ocean, in a way that respects our planet and allows local communities to flourish." Mowi's sustainability strategy is around people and planet and their most important communication is on providing healthy food for a growing population, resource efficiency, exploited resources, aging population, and climate change. They disclose a visualization of their strategy in their sustainability report and the value they create to stakeholders. They disclose stakeholder materiality assessment and have listed their priorities on planet product, people, and profit. They still very strongly communicate that they are profit-driven rather than purpose-driven. Their material environmental impacts are climate-friendly production, preventing fish escapes, responsible sea lice management, responsible use of medicines and chemicals, efficient and sustainable fish feed, and ensuring fish health and welfare. They also use SDGs for communication. They have a clear overview of commitments and **commit to reduce scope 1 and 2 emissions by 50.6%, reduce scope 3 emissions by 27.5%, and reduce scope 3 FLAG (Forest, Land and Agriculture) GHG emissions by 33.3%, all by 2030.** They have commitments to the plastic consumption being recyclable, reusable, or compostable (not-recommended by industry). They have other planet commitments in their overview. **They do not disclose a transition plan to climate zero.** Innovation is a central part of their sustainability report.
https://mowi.com/wp-content/uploads/2024/05/240512-Mowi-Sustainability-Strategy.pdf
Gender equality in management (F/M):
EMT – 2/7 (22%); BoD – 4/5 (44%)

Other corporations that are important in the food value chain are the retailers as they source as well as communicate with the consumers. Larger retailers in the Nordics and other companies that are interesting to look at are mentioned here:

Company	Industry	Country	Link
ICA	Retail	Sweden	https://www.ica.se/
COOP	Cooperative retailer in the Nordics owned by customers, locally and sourcing collectively	Denmark, Norway, Sweden, Finland	https://coop.dk/ https://www.coop.no/ https://www.coop.se/ https://www.cooptrading.com/
Rema1000	Retail operating in Norway and Denmark	Norway	https://www.rema.no/
Carlsberg	Large brewery operating globally	Denmark	https://www.carlsberg.com/en/
Eskja	Fishery	Iceland	https://eskja.is/en/
Islandsfisk	Fishery from Iceland	Sweden	https://islandsfisk.se/en/

Food producers in Finland and Iceland are more locally based and these countries do not have the same kind of large industries within food production as the rest of the Nordic countries.

COOP is a cooperative locally in the Nordic countries that is owned by the customers in every country. Recently 50% of COOP Denmark was taken over by OK (gasoline and diesel provider) to counter bankrupcy. Originally the focus of COOP was on better food and sustainable production, but they are lacking behind in the transition and are as the rest of the retailers and supermarkets not strongly focused on engaging and driving the transition together with their customers.

In the discussion on sustainability and healthy diets, it is important to mention the largest global corporation in the Nordics. Novo Nordisk at a point reached the 15th largest corporation globally and the only Nordic company on the global top 100 measured by market cap.

2.2.11 Novo Nordisk

Novo Nordisk
Novo Nordisk is privately owned and listed in Denmark. The Novo Foundation owns A- and B-shares amounting to approx. 28% of share capital and approx. 77% of the votes.
https://www.novonordisk.dk/
Revenue (2023): 31,134 billlion euros. Number of employees: 64,300

Product and business model:
Novo Nordisk is a medico corporation and one of three largest global insulin providers. Today Novo Nordisk (NN) also produces other medical products and has recently launched Wegovy a new medicine used to reduce calorie diet and increase physical activity to cure either obesity or overweight. This has become such a market success and has increased the market cap of Novo Nordisk to one of the largest corporations globally, continuously expanding their production.

Sustainability:
Purpose: To drive change to defeat serious chronic diseases, built upon our heritage in diabetes. Sustainability approach is communicated as: Doing more with less. Novo Nordisk has for many years been in the forefront of sustainable global corporations, very early introducing shared values as an integral part of its corporate strategy. They have historically introduced educational materials for doctors and paramedics to prevent and treat diabetes, also in less developed regions. They have support programs to provide cheap insulin to the poorest people. With the new product Wegovy they widen their market potential enormously as obesity and overweight have become one of the largest health problems globally. More people still die from obesity than from hunger; this will change with climate change as, in large parts of the planet, especially in the tropic and subtropic areas, food production will be significantly threatened and cause draught, famine, and migration flows. The size of the global GLP-1 (the active substance in Wegovy and Ozempic) analogues market was estimated to be worth $47.4 billion in 2024 and to reach $471.1 billion by 2032. The effects of this medicine will change the consumption of sweets and food patterns as well as the medical budgets of nations. There should be an ethical discussion on providing medicine to cure obesity and overweight rather than changing overconsumption of unhealthy diets on a planet that cannot provide for the existing or future population unless consumption is changed dramatically. As well as the issue that a few private enterprises such as Novo Nordisk and Eli Lilly profit from this new medicine to an extent never seen before globally.

Looking at Novo Nordisk and ESG, there is no doubt that they have one of the world's most impressive and extensive sustainability communications and reports. The overall vision of NN: **Add value to society and to our future business**. To achieve this ambition, we strive to do business in a financially, environmentally, and socially responsible way." They communicate in the framework of ESG and have a public accessible portal to communicate and disclose their ESG commitments, targets, and data. They have a Sustainability Advisory Council seated with external global experts. There is no doubt that NN is a corporation to be inspired by in their sustainability work, commitments, and communications. NN has a target of **reaching zero CO_2 emissions from operations and transportation by 2030 (scope 1 + 2)**. Scope 3 emission is by far the largest emission and NN discloses that its suppliers' activities account for the majority of the total CO_{2e} emissions (98%) in 2023. Their target is that all goods and services from suppliers will be based on 100% sourced renewable power by 2030, but this does not make NN climate neutral as they also need to work on their downstream emissions, especially from the packaging materials that they sell. **NN has a target of net-zero CO_2 emissions in 2045 across their full value chain (scope1 + 2 + 3)**. Here minimizing the use of virgin fossil-based plastic, reducing the consumption of plastic, and changing the plastics to non-virgin fossil types will cover the largest part of the emissions. This seems a little unambitious with the outreach and financial power they possess and invest in sustainability. A large part of their value creation is collected by the major owner the Novo Nordisk Foundation—a foundation of the size of the Melinda and Bill Gates Foundation. These two foundations have joined forces in addressing some of the large planetary and social problems through support programs, which is very valuable and an example to follow by other global corporations including the dominating tech corporations. The democratic control of these foundations should be discussed, as a very large share of global value is directed and administrated by very few appointed businesspeople. No doubt the foundations operate transparently and with very responsible intentions but outside the control of societies, not only the local societies where they are registered, but also outside the global society where they harvest the profit that generates these enormous sums of money.
https://www.novonordisk.com/sustainable-business/esg-portal.html

Gender equality in management (F/M):
EMT – 2/9 (18%); BoD – 6/6 (50%). Noticeable: 3 out of 4 members of the board elected by employees are female.

Generally, the corporates in the food sector driving the transition to sustainable land use, ecosystems, and healthy diets are generally less prepared and have sustainability less integrated into the core business, compared with the energy sector corporations, except for Novo Nordisk. Scope 3 emissions and circular economy are less in the scope of their commitments. In Denmark, the largest GHG emissions come from the agricultural sector and the sector has a huge impact on the marine and freshwater ecosystems of the Nordics including the North Sea, and this is not strongly reflected in their approach to sustainability, yet—generally (Fig. 2.4).

2.3   Sustainable Transport

Sustainable transport is an important issue in the Nordics as many of the goods manufactured and consumed here are transported by road in diesel trucks coming from the south, landed in harbors in Germany and Holland, and freighted to the Nordics. The Nordic Region historically held a strong infrastructure of transport by ocean and rivers, as well as by train. But as in other parts of the world the transportation of humans and goods has moved to roads making transportation very energy

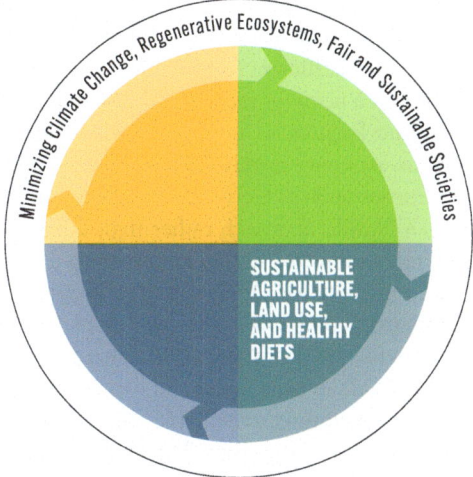

Re-establishment of natural habitats to rebuild biodiversity globally.

More balanced diets lead to healthier and more affordable food, minimizing obesity and other diseases that strain humans and our economy today.

A closer connection between food producers and food consumers will accelerate the demands for more sustainable and local production since we will again know how animals are treated and how pesticides are added.

Urban Farming to green the cities, make food production come closer to people and create an understanding of methods and influence on biodiversity. Taking the cars out of the Cities will leave space for Urban Farming since we will need much less space for transport.

Intelligent use of bi-products and organic waste in a cascade before bio-composted. There are huge potentials in extraction proteins, fibers, and lipids from organic waste and there is a huge potential in using organic bi-products and waste for feed into the production of new types of meets and proteins from worms, insects, fungi, etc.

Fig. 2.4 Climate Nexus – Sustainable agriculture, land use, and healthy diets. The Climate Nexus is a concept described in the book: The Great Transition to a Green and Circular Economy. Gitte Haar. 2024 and visualizes a holistic approach to implementing sustainability

inefficient, and road traffic has created jams, accidents, and inefficiency. The combustion engine is a heavy and inefficient energy consumer compared to trains and ships. Freight by road trucks is an inefficient means of transport as more than 50% of the trucks drive empty on the roads and the system is very vulnerable. Road trucks take up a significant part of the road capacity and are also to a large extent involved in road accidents often with lethal outcomes.

The prices of transportation by road are often perceived cheaper than by rail in many countries, for both people and goods. When looking at the total cost of ownership (TCO) and taking load potential into account, rail is often cheaper than trucks,[2] but less flexible. The costs of the externalities are not included in any of the

---

[2] https://www.rsilogistics.com/blog/comparing-the-costs-of-rail-shipping-vs-truck/

scenarios, as for example the price on climate change, pollution, and noise. When this is considered, rail will far outweigh the prices of road for both people and goods.

Generally, the transition to inefficient road traffic of goods has evolved over the last 20–30 years, and there is a need for retrofitting and revitalization of the logistic infrastructure in the Nordic Region, as in many parts of the world. Sweden has announced targets and investments in developing the rail infrastructure together with other countries in Europe such as Germany. The development and shift back to rails and ships requires a Nordic approach rather than a national approach as the infrastructures must be interconnected across countries to transform to sustainable transport and making cities livable for people rather than cars. Denmark holds some of the world's largest logistic companies with Mærsk and DSV, now taking over another large logistic provider in Germany (DB Schenker) originally owned by Deutsche Bahn—with the German state is the real owner. It is interesting to see the vision for this in relation to developing sustainable means of transport on rail and by ship and excluding road traffic from the cities. The Nordics hold no united approach to transform the region to sustainable transport at this moment, and the transition to sustainability here is behind what is seen in France, Germany, and China where the goal is to connect all larger cities by fast speed trains.

The transition to sustainable transport in all three scopes is challenging from a technological and implementation point of view. It is not solved solely by the electrification of vehicles as many do now. It requires shift from roads toward shared transportation by rail and ship, legislation, infrastructure, inclusion of city road traffic policies, and pricing of corban emission, pollution, and roads. The impacts from aviation per unit transported are huge and the transition here is complicated as the airplanes are not easily electrified or substituted with other fuel types. Not even the Nordic countries are able to find common ground to showcase a sustainable transition even though the Scandinavian states owned airlines. The international airlines in the Nordics are reviewed here:

2.3.1   SAS Group

SAS Group
Today SAS is mainly in private hands with the Danish state holding 25.8% of the shares and the rest owned by KLM/Air France and Castlelake, an American investment company from the recent restructuring of the company in 2023/24. SAS has just been delisted from the stock exchange to restructure the company hopefully also including finding new types of fuels.
https://www.sasgroup.net
Revenue (2023): 3.7 billion euros. Number of employees: 10,000+

Product and business model:
Scandinavian Airline System provides airline services with headquarters in Stockholm and main hubs in the capitals of the three Scandinavian countries operating approx. 24 million passengers with global destinations. In addition to flight operations, SAS offers ground handling services, technical maintenance, and air cargo services.

Sustainability:
The vision: "SAS aims to be the driving force in sustainable aviation and in the transition toward net zero emissions. We are continuously reducing our carbon emissions through using more sustainable aviation fuel, investing in new fuel-efficient aircraft and technology innovation together with partners—thereby contributing towards the industry **target of net zero CO_2 emissions by 2050**." Rather unambitious with lack of sub-targets.
The sustainability paragraph in the annual report of SAS is very generic with few specific targets. They include a materiality assessment based on the old methodology of stakeholder engagement, not the recent requirement of double materiality assessment based on ESG impacts versus financial potentials and risks.
SAS being financially challenged holds the opportunity to put sustainability and zero emissions as an innovative power to create competitive advantage over the larger airlines — if they can or intend to do so is not clear.
https://www.sasgroup.net/files/documents/financial-reports/2023/SAS_Annual_and_sustainability_report_FY_2023.pdf

Gender equality in management (F/M):
EMT − 2/6 (25%); BoD − 2/8 (20%).

2.3.2   Norwegian Air Shuttle

Norwegian Air Shuttle
A Norwegian privately owned company listed in Norway and the Norwegian state which holds 8% of shares. The company was started by a private person and has been supported by the state.
https://www.norwegian.com
Revenue (2023): 2.2 billion euros. Number of employees: 5500

Product and business model:
Norwegian Air Shuttle is the third largest low-cost carrier in Europe.

Sustainability:
The vision is: "dedicated to fostering enduring value for all stakeholders, encompassing customers, shareholders, employees, and society as a whole." In pursuit of this goal, the company prioritizes "a robust safety culture and conducts operations with responsible, sustainable business practices grounded in sound environmental, social, and governance (ESG) principles." NAS discloses an integrated annual report with a section on ESG. Very general and text heavy communication on ESG that does not give an inclusive or ambitious impression. They commit to SDG#3, 5, 7, 8 9, 13, and 17 and display an overall roadmap on these SDGs, but with no targets. The text discloses some targets and an explanation on **climate target of 45% reduction in relative carbon intensity by 2030**. Zero food waste by 2030. They disclose their emissions in the 3 scopes and 80% of the emissions come from fuel in scope 1 and the rest from scope 3. They offer customers to offset their climate impacts but does not clearly communicate net-zero emissions targets. This is a poor and uncreative way of communicating sustainability. They have a web interface on sustainable aviation that communicates. They participate in fossil-free fuel trials, but generally they leave an impression of superficial and unambitious sustainability approach.
https://www.norwegian.com/uk/about/experience-us/responsible-travel/

Gender equality in management (F/M):
2/6 (25%); BoD − 3/5 (38%).

Some other large transport corporations operating shipping, logistics, and freight in the Nordics are reviewed:

2.3.3 Mærsk

Mærsk
Mærsk is a privately owned company listed in Denmark with the majority of shares owned by Foundations and Holding company dominated by the founder's family A.P. Møller.
https://www.maersk.com
Revenue (2023): 46 billion euros. Number of employees: 106,000

Product and business model:
Mærsk is an integrated container and logistics company and one of the globally largest providers of shipping and freight. Mærsk also owns and operate harbors around the world. They divide their operation into transportation services, supply chain and logistics, and digital solutions.

Sustainability:
The vision is: "Integrating the world for a sustainable future." Their **climate target is net zero by 2040**, with a roadmap approved by SBTi. Mærsk has for many years promoted sustainability especially within their shipping and container business and has developed low impact shipping by enlarging ships, minimizing energy consumption, and now pursuing inputs from non-fossil energy sources. They launched a full green fuel ship in 2023. Mærsk sees methanol as the best solution for now, meaning fuel from renewable sources or biological (organic) sources. They team up with some very large producers of energy in scalable solution for green methanol. It is mysterious that they do not include sails as part of their solution to climate-neutral shipping, as the wind on oceans has moved people and goods for thousand years already. A combination of technologies is always the solution to the transition. Mærsk has also historically been strong in the communication on sustainability and has an integrated sustainability report and links on ESG communications on the web. They communicate in the ESG framework, and the E is mainly covered by climate change, environment and ecosystems, and responsible ship recycling. Their social and governance communication fairly aligns with the CSRD. They do not specifically address the business impacts of a circular economy, which means more local production, more repair and maintenance, and more take-back of used products. The circular economy is also a new business imperative for the shipping and logistics industry.
Another issue is that Mærsk is an integrated provider of transportation and freight, and that part of the value chain on land has the heaviest environmental and social impact. Mærsk has an increased focus on growing their business in the developing economies and the growth countries especially in Africa. Generally, global freight and shipping is challenged by corruption and bribery and Mærsk is addressing this as one of their values to avoid, but cases still occur around the world. Their communication and reporting on sustainability has been awarded and it is a good inspiration for others, even if it leaves a doubt whether the largest business issue from the great transition to a green and circular economy has been evaluated in relation to their overall strategy apart from focusing on the developing regions.
https://www.maersk.com/sustainability

Gender equality in management (F/M):
EMT – 4/10 (29%); BoD – 3/7 (30%).

2.3.4 DSV

DSV
DSV is a privately owned corporation listed in Denmark.
(https://www.dsv.com/)
Revenue (2023): 20.2 billion euros. Number of employees: 75,000

Product and business model:
DSV is a global provider of transport and logistics solutions within road transport, air and sea freight, logistics, and warehousing. DSV is a B2B operator and forwarding agent and holds an asset light business model and does not own the trucks, airplanes, or ships they operate. The total fleet amounts to over 15.000 trucks, and they are specialized in FMCG business.

Sustainability:
The vision is: "Towards a more sustainable future." DSV target is **net-zero emissions across all operations before 2050**. They do not display a transition plan, and the net-zero target seems unambitious. They hold a decent sustainability communication and have just announced the delivery of 300 electric trucks from Volvo as part of their strategy on sustainable road transport. Their focus is around facilitating solutions and documentation for customers now. They promote green logistics with specific decarbonization solutions ranging from CO_2 reporting to strategic supply chain optimization, sustainable storage solutions, sustainable fuels, and carbon offsetting. Online, DSV display an interactive map on 139 Sustainability Impact Initiatives, globally, and these are illustrated with the SDG icons and on CO_2 reduction, employee volunteering, financial donations, and in-kind services with a significant majority of social initiatives. This way of communicating is creative, authentic, and inspirational. https://www.dsv.com/en-gb/sustainability-esg

Gender equality in management (F/M):
EMT – 0/3 (0%); BoD – 3/5 (38%)

2.3.5 Nordicon

Nordicon
Originally a Swedish owned corporation, now owned by Allcargo Logistics Ltd.
Revenue (2023): billion euros. Number of employees: not disclosed separately

Product and business model:
Nordicon is a provider of ocean freight, rail service, domestic trucking, and warehousing operating in the Nordic countries. Nordicon specializes in ocean and rail freight and can handle containerized cargo in full containers. Nordicon has just announced that they offer instant access to renewable fuels.

Sustainability:
Vision: Nordicon should be the most preferred and leading neutral consolidator in the Nordic region with a business practice based on the standards of ESG. Their take on sustainability is integrating ESG into ECU world. By exploring environmental practices, increasing environmental efficiency, transitioning to renewable and clearer sources of energy from non-renewable when not feasible, and exploring technical solutions to reduce environmental footprints. **Stating no commitments or targets on climate neutrality** or environmental impacts it becomes very unclear what the actual transition to sustainability is about.
Interesting announcement on their frontpage: "As part of our ongoing commitment to reducing carbon footprints and promoting sustainable practices, Nordicon can now offer our clients a seamless way to opt for renewable fuels. As a part of ECU Worldwide Group, Nordicon is dedicated to revolutionizing sustainable logistics and building a more sustainable future, one shipment at a time. This new product is a significant part of our group's initiative, GreenwayZ. The product is now available in Sweden. Expansions to Denmark, Finland, and Norway to follow later this year." This may be a significant introduction of fuels dependent on the level at which they are able to substitute the fuel. Full substitution of fossil is significant; part substitution is industry standard.
https://www.nordic-on.com/why-nordicon/sustainability/

Gender equality in management (F/M):
Not disclosed

Other important logistic providers in the Nordics relevant to review are listed here:

Company	Industry	Country	Link
DFDS	Shipping and logistics provider in Europe	Danmark	https://www.dfds.com/
FREJA Transport & Logistics A/S	A full-service shipping and logistic operator within transport on road, ocean, and air	Denmark	https://en.freja.com/
Green Cargo	Climate- and cost-effective logistics for Swedish businesses	Sweden	https://www.greencargo.com/
NTG Nordic Transport Group	Transport company offering tailor-made road transport to and from all the Nordic countries	Denmark	https://ntgnordic.dk/?lang=en
Scan Global Logistics	A full-service domestic and international freight forwarder	Denmark	https://www.scangl.com
Scandinavian Express	A transport company operating primarily in Scandinavian countries—mainly Norway, Sweden, and Poland	Sweden	https://www.scandinavian.com.
Leman	A one-stop-shop for supply chain and logistic solutions	Denmark	https://leman.com/

Rails and railway services are mostly run by the states in the Nordics with Danske Statsbaner (DSB.dk), Statens Järnvägar (SJ.se), Bane Nor (banenor.no), and VR Group (vrgroup.fi). They operate across the Nordics. Iceland has no public rails due to the non-dense population and the rough landscape. The railway systems in the Nordics are very widespread and have been in place for 160 years; nevertheless, this shared mode of transport is losing passengers these years, and is outcompeted by trucks, cars, and airlines. Price levels vary across the Nordics with Denmark and to

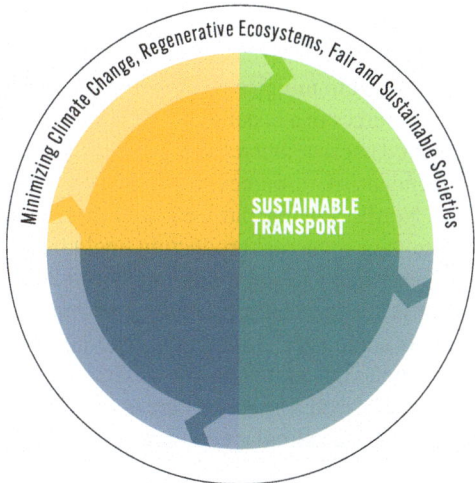

Redesign of collective, public, electrified transport in cities contributing to smart, livable, and sustainable cities.

Minimize particle and noise pollution that causes healthcare issues for millions of people.

Solve congestion of traffic in cities to make space for increasing urbanization and make cities livable with recreative areas, local food production and biodiversity.

Fig. 2.5   Climate Nexus—Sustainable transport. The Climate Nexus is a concept described in the book: The Great Transition to a Green and Circular Economy. Gitte Haar. 2024 and visualizes a holistic approach to implementing sustainability

some extent Sweden being expensive, and Norway and especially Finland have affordable and efficient train systems, also including fast trains. Denmark has not yet introduced fast trains. In Norway, they market their fast trains to to compete with airlines between the cities.

Nevertheless, there lacks a Nordic consolidated strategy for sustainable transportation that will connect the cities and be central in sustainable transportation between and within cities to minimize environmental impacts and making cities livable. It is important to note that the construction and maintenance of railway systems has less environmental impacts than that of individual freight and transport. Not only because railways are shared and efficient they are the most eco-friendly means of transport on land (Fig. 2.5).

An overall assessment of sustainable transport in the Nordics is that the potential to transform here is huge but the roadmap and shared visions across the Nordics are lacking for all means of transport. The states are privatizing the airlines, and the road transport may or may not lead to a fast transition to sustainable transportation. In all cases, it is necessary for the political levels to start gathering and demanding for the transition to happen and to invest massively in shared transportation on rails and by ship, also locally.

2.4   Circular Economy

Circular economy is important to mitigate climate impacts because mining and the continuous use of virgin resources in a linear economy based on take-make-waste are energy-consuming compared to reuse and recycling. Circular economy is also important in becoming independent of resources imported from outside the European region. Historically the Nordic Region has delivered significant amounts of raw materials, especially metals for industrial production, including iron ore, aluminum, cobber, oil, and silver. Today many of the mines are closed because they have been exhausted or the mining is too expensive compared to mining in the developing countries. Greenland holds some very critical raw materials that will become very attractive when the ice is melted. Now the people of Greenland have decided not to mine these raw materials and instead conserve nature as the main value for Greenland.

The important objectives of the circular economy are:

- To minimize energy consumption from harvesting and processing virgin resources by reuse and recycling.
- To become independent of virgin raw materials by harvesting from the existing resources
- To minimize production of waste as waste and land fill cause environmental disasters and GHG emissions
- To retain critical materials and economic value that have become scares
- To minimize pressure on natural ecosystems
- To minimize ESG impacts from linear consumption of take-make-waste, as well as minimizing economy waste

Implementing a circular economy is a transition to sustainable and responsible consumption patterns. In the Nordic region, a large part of the climate impact originates from the sourcing and consumption of goods manufactured outside the region and outside the EU. Above 80% of the climate impacts of most companies in the Nordics in various industries are in scope 3—upstream and downstream by importing products and parts. Combined with linear consumption causing waste and loss of values, the environmental impacts are huge. Overconsumption in the Nordics is a big part of the environmental impacts, and the Nordics hold some of the most significant companies in the linear economy, as H&M, Bestseller, IKEA, JYSK, and others. A transition to responsible consumption in the Nordics is an important part of lowering the climate impacts as well as stopping the huge production of waste that is happening here. The waste production per capita is shown in Table 2.1.[3]

As seen from Table 2.1 the Nordics are high consumers and high producers of waste only exceeded by the USA, the country that generates most waste globally per capita. In the EU, the yearly average amounts to 513 kg/capita and the Nordic citizens, except Sweden, are high contributors to this high generation of waste. Sweden is an old industrial country with traditions for recycling and containing values,

---

[3] https://www.statista.com/statistics/789638/production-waste-tons-by-inhabitant-union-european/

Table 2.1 Waste statistic per capita in 2022

Country	Kg waste/capital (2022)
EU	513
Denmark	786
Norway	768
Iceland	659
Finland	630
Sweden	475
USA	835

which also keeps the waste generation low compared to other countries, and very low compared to the rest of the Nordics. There is a need to reuse and recycle much more for economic as well as environmental reasons.

The EU as well as the Nordics are introducing regulation on containing critical raw material necessary for the local production and consumption and to become less dependent on imported goods and materials. The EU Green Deal, the Circular Economy Action Plan (CEAP[4]), and the EU Critical Raw Materials Act (CRMA[5]) have the overall goal to become independent of virgin resources by creating an economy based on the efficient use of resources meaning harvest from what we already have, namely the waste.

National strategies and action plans for a transition to circular economy are in place in the EU strongly supported by the regulation toward companies and their products. In the Nordics all the countries, except Denmark, have a national strategy on circular economy.

The national strategies for Norway, Sweden, and Finland, as well as the strategy of the Nordic Council on circular economy, are very much aligned. Strategies before 2020, where the EU Green Deal was released, are considered outdated and then Denmark (2018) does not hold a national politically approved strategy for a circular economy. The overall Nordic strategy for introducing the circular economy is illustrated in Fig. 2.6.

It is very important that the Nordics work together on introducing a circular economy as a region because the market of each country is too small to drive profitable business and investments into the circular economy. The transition to a circular economy is driven by the companies, which is why this Case Collection includes showcases on companies transformation to a circular economy in Chap. 3.

Nordic Innovation under Nordic Council has founded the initiative to facilitate and share knowledge on the transition to a circular economy, called Nordic Circular Hotspot (NCH). NCH host partnerships, summits, events, and knowledge sharing platforms and are also inviting the Baltic countries to participate in the activities. See: https://nordiccircularhotspot.org.

---

[4] https://environment.ec.europa.eu/strategy/circular-economy-action-plan_en

[5] https://single-market-economy.ec.europa.eu/sectors/raw-materials/areas-specific-interest/critical-raw-materials/critical-raw-materials-act_en

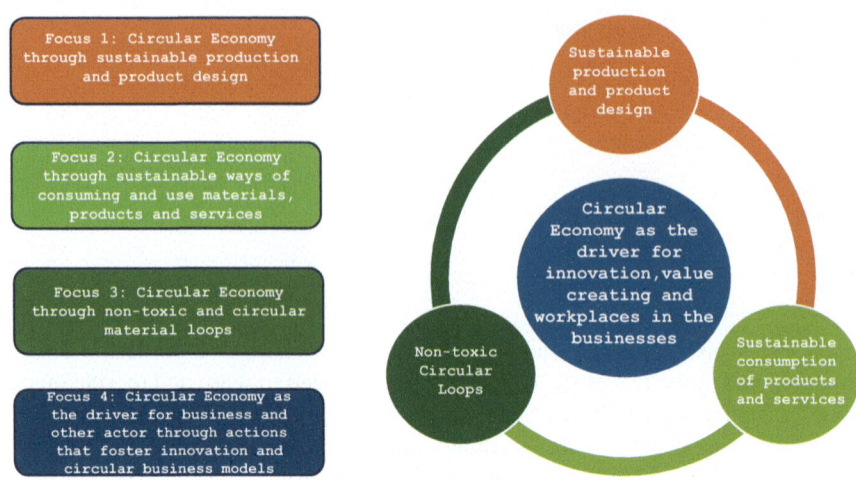

Fig. 2.6 CE strategies in the Nordics. The strategies for implementing a circular economy are very similar in the countries that have national strategies, such as Norway, Finland, and Sweden

An overview of the Nordic Circular Policies and Regulations is available and includes not only circularity policies but also policies on energy targets transition and others. See: https://nordiccircularhotspot.org/publications/nordic-circular-policies-and-regulations

Country	Title	Link
Norway	National strategy for a green and circular economy (Nasjonal strategi for ein grøn, sirkulær økonomi)	Nasjonal strategi for ein grøn,sirkulær økonomi (regjeringen.no)
Sweden	Circular Economy—Action plan for the transition of Sweden (Cirkulär ekonomi—Handlingsplan för omställning av Sverige)	https://www.regeringen.se/contentassets/4875d d887fd34edabd8c1d928a04f7ba/cirkular-ekonomi-handlingsplan-for-omstallning-av-sverige.pdf
Finland	Strategic program to promote a circular economy	https://ym.fi/en/strategic-programme-to-promote-a-circular-economy
Iceland	Icelandic waste law also called *The Circular Law* Building up the Circular Economy in *Blue* Iceland (Ocean Cluster)	https://urgangur.is/a-new-law-on-waste-takes-effect-in-2023-what-will-change/ Building up the Circular Economy in Iceland - Íslenski sjávarklasinn (sjavarklasinn.is)
Denmark	Danish Industry (industry body)	https://www.danskindustri.dk/politik-og-analyser/dis-politiske-udspil/di-politik-for-cirkular-okonomi/
Nordic Council	The Working Group for Circular Economy (NCE)—2019 to 2024. Needs update	https://www.norden.org/en/information/nordic-working-group-circular-economy-nce

National strategies for Denmark and Iceland and an update of strategy from the Nordic Council are still lacking. Finland seems to be moving fast toward a circular economy in the Nordic Region and good cases are available here. Finland has also included taxation as a tool in their national strategy. Norway and Sweden are working top-down focusing on the recycling and waste handling industries, whereas Finland is engaging in the transition through innovation and investments in the companies that implement a circular economy into their business models.

All the Nordic countries are now (2024) releasing public funds (grants) for the circular transition, and it will be interesting to see if this will accelerate the transition, and must be strongly supported by political ambitions and focus. A national strategy on circular economy together with the large corporations to take responsibility for the circular transition according to requirements in CSRD is necessary. This will also push the SMEs into the circular transition through supplier requirements and investments. Nevertheless, now it seems as if the circular innovation is happening mostly in the SMEs as described later in this book.

First and foremost, the EU product regulation (EU Sustainable Production Regulation—ESPR) imbeds circular economy by imposing longevity of products, right to repair and maintain, and an extended producer responsibility for the full value chain making manufacturers and distributors in the EU market responsible for disposal into a full reuse and recycling system. Now, it is left to hope for rapid implementation. More details on this can be found in another book by the same author (Haar, The Great Transition to a Green and Circular Economy. Climate Nexus and Sustainability, 2024a).

Every year a World Circular Economic Forum is held and learnings from WCEF 2023 can be found in the booklet: Nextcloud.[6] From this booklet are highlighted important initiatives to take notice of and to learn from in the Nordics on Circular Economy (May 2024):

Initiative	Link	Country
CircIT and Haga Initiative	www.hagainitiativet.se/wp-content/ uploads/2023/12/231212-Co-benefits-of-Circular-Economy-in-the-Nordics-1.pdf	Sweden
RE: source—project database	Projektdatabas med alla projekt—RE:Source (resource-sip.se)	Sweden
CradleNet	https://www.cradlenet.se/medlemsintervjuer Möt Cradlenets medlemmar—Cradlenet - Accelerating Circular Economy	Sweden
Sitra	https://www.sitra.fi/en/ articles/41-pioneering-fnnish-circular-economy-companies/	Finland
KiSu—hub for Circular Economy	https://kiertotaloussuomi.fi/en/	Finland
Nordic Circular Economy Playbook (Nordic Innovation/ Nordic Council)	Nordic Circular Economy Playbook	Nordic Innovation	Nordic

---

[6] Nextcloud (webo.hosting)

When evaluating the transition to a circular economy in the Nordics it is important to look at the materials, the value chains, and the recycling industry.

Challenged Industries and Materials in a Circular Economy
Some industries and materials flows are more challenged and necessary to transform to implement the future circular economy. Important materials in the Nordics where no or little recycling at scale is in place are plastic and textiles. Large amounts of single-use plastic and textiles are put on the markets due to the overconsumption patterns and cheap prices required by the market and possible by the cheap sourcing from Asia. General recycling and reusing of materials and goods, as well as building infrastructure and loops to facilitate recycling and reuse, is in necessary to implement along the value chains. Metals and glass have been recycled over decades in this region, but still the share of recycled materials of all types of materials needs to increase. Building materials are the largest share of all waste not only in the Nordics but also in the EU. Up to 40% of all waste come from the contruction industry. Significant focus is on reuse, recycling, and prolonging the lifespan of building materials generally in the EU to stop leaching important resources. It is importnat to come back to an economy dependent on recycling and reuse rather than single use of products with a short lifespan.

With overconsumption comes a huge amount of overproduction that never meets the consumer, especially in the building industry and the textile industry but also in food production. This is becoming more and more outspoken due to the low production costs in Asia and the lack of sustainable demands in the consuming countries. We need to move into a circular economy as illustrated in Fig. 2.7 (Haar, The Great Transition to a Green and Circular Economy. Climate Nexus and Sustainability, 2024a).

The purpose of organizing the full value chains in a circular economy is to become independent of virgin inputs and to avoid creating waste through the two circles illustrated in Fig. 2.7. It is important to:

- Design products for reuse, recycling, and disassembly.
- Design products from mono-materials or components of mono-materials that can be disassembled.
- Use and reuse the products as designed for as long as possible through maintenance and repair.
- Sorting and collection of the materials for recycling at highest possible value, to create infinite loops of materials for future products.

The most important drivers toward a circular economy are:

- Access to well-sorted and well-defined materials at the same levels as earlier known from virgin materials through material banks
- New circular business models that ensure new consumption patterns and take-back of products and materials

All this is built into the EU regulation broadly on company level and on product level as mentioned above.

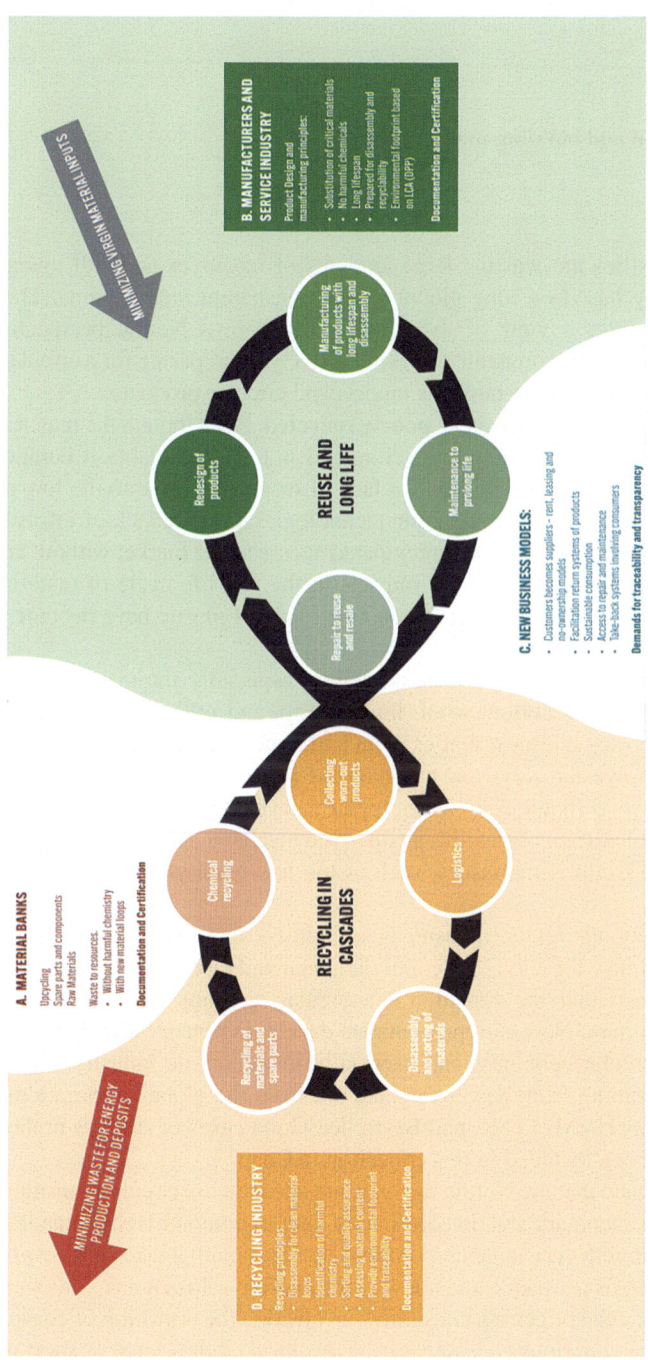

Fig. 2.7 Circular Economy. The circular economy is a redesign of the value chains and foster new collaborations and new business models based on product loops of reused products (green) and recycled materials (yellow)

In the Nordics as well as generally, the following industries are challenged in the transition to a circular economy:

• Plastics
• Textile
• Construction and building materials
• Packaging
• Recycling

These industries are wasting large amounts of resources and still operate in a linear economy based on long global and untransparent value chains. This Case Collection provides a view of these industries from a Nordic perspective and includes some cases of companies working actively in preparing for a circular economy by developing circular business models and circular products.

As in all regions, **plastic is very poorly recycled.** Even though the material itself has high potential for reuse and recycling and is fully recyclable if composed of known mono-polymers. The reason for lack of recycling is due to the low price of plastic, the lack of profitable recycling technologies, and the lack of infrastructure for recycling. The problem is the unregulated access to the market without accounting for disposal and pricing of the impacts in the full life cycle of materials and product. Then cheap products and packaging made of plastic have become single-use rather than recycled.

Textiles have an old history of recycling, especially the textiles made from natural fibers, such as cotton, wool, linen, hemp, and other natural fibers. Today recycling of textiles is almost non-existent and almost all textiles contain synthetic fibers such as polyester, acryl, and nylon mixed with the natural fibers making recycling at fiber level difficult. New technologies must be developed, tested, and implemented to be able to recycle textiles again. Even mechanical recycling at fabric level is difficult due to the same issues, such as too low quality and additives difficult to assess.

Unfortunately, the textile industry is an industry that has been communicating about sustainability for a long time, but the green and circular transition is not taking place. In particular, the transition to a circular economy is very far away in the textile industry here. Now entrepreneurs are developing business models and products for the circular economy, and especially in the textile industry, change and disruption is coming from new actors and not from the global corporations in the Nordics, such as H&M, JYSK, and Bestseller. Good cases of startups in the textile industry have been included in this collection in Chap. 3.

It is unfortunate that the Nordic region is not ahead in the circular transition of the textile industry. Denmark and Sweden have a long tradition of textile manufacturing and design, and with very large corporations that have built a take-make-waste business on sourcing from opaque and not very responsible value chains. This should not belittle the challenge of getting consumers involved in the transition of consumption of cloth, as even conscious consumers still buy cheap clothes for very short use.

It seems as if Finland is moving on the circular transition in the textile industry. Neutral (https://neutral.com/) is one of the few Danish textile companies that works seriously with sustainability and is open about the absence of the circular economy in the industry.

Really (https://www.kvadrat.dk/en/kvadrat-really) is an entrepreneurial company, now owned by a fabric manufacturer that was born circular and utilizes textile fibers that were previously burned as waste and which today are turned into a resource, in the form of new materials used for furniture and other exciting products.

In total, **packaging waste** accounts for one-third of all household waste, and this is because packaging materials are also becoming more and more single-use. As an example, old-time reuse logistic packaging materials have been replaced with card boxes made of recycled fibers but for single-use. Single-use plastic is also a large fraction of the waste generated from packaging materials especially for food packaging and is causing not only a huge leak of fossil-based resources but also pollutes nature and oceans to an extent that politicians have been forced to react. Now legislation is imposed on the fillers of this packaging material as well as on the plastic industry itself in the EU to move into a circular economy.

Collection and sorting of waste from industries and household is extensive and has a history in the Nordics compared to other regions. The Nordics together with the Baltics are implementing sorting standards to ensure well-defined and sorted fractions of resources which increases the recycling potential. EU has passed regulation to ensure uniform sorting of waste/resources to ensure sufficient volume for recycling of material loops. The recycling industry is moving fast into a circular economy, but reuse and recycling still lacks investments in technologies, infrastructure, and economic incentives. As the virgin products and materials are still too cheap. This will be regulated with the new EU Circular Economy Action Plan (CEAP) and the new EU Sustainable Product Regulation (ESPR).

The recycling industry seems to be very innovative and an industry that has caught the eye of the landscape of the circular economy. For decades, the recycling industry has worked with the extraction of resources from waste and upcycled materials from waste. Many of the recyclers emerged in times of resource scarcity and at a time when building materials were reused and recycled. In times when demolition companies paid to demolish buildings and sold the used building products at the highest possible quality to gain high prices. Today the business model and cash flow model are the opposite; today they are paid to demolish a building and get rid of waste. This must be reversed back again. The innovation here is great and some of the top material experts sit in the recycling companies. There is no doubt that demolition industry will own a larger part of the value chain in the construction industry, and in many other industries. They become the future material banks and providers of recycled and reused materials. The demolitions and building material manufacturers must meet and collaborate to create the material loops of the future.

The linear value chain of the construction industry creates massive volumes of waste as products are designed for much shorter longevity than just decades ago and thereby the industry has created business models based on standardized products being replaced frequently rather than 50 years ago where most building materials were repaired and reused. This linear, take-make-waste model, as in other industries, also creates huge amounts of waste from construction sites of products that are never installed due to mismanagement and mis-design or just because it is easier to discard an open batch of insolation rather than rearranging it to the next construction site.

In Chap. 3, recyclers of plastic, textiles, and building materials are included to showcase the innovative solutions. Companies that are important in scaling recycling of plastic, textiles, and other materials into a circular economy.

Some of the large Nordic corporations that have lived from linear business models and the take-make-waste from long, transparent supply chains are reviewed here:

2.4.1   IKEA Group

IKEA Group

Originally a Swedish corporation privately owned by Ingvar Kamprad. Today most of the stores are owned by a private Dutch company Ingka Holding. This company is fully owned by a Dutch foundation, Stichting Ingka.

https://www.ikea.com/

IKEA Group also includes IKEA Foundation: https://ikeafoundation.org/ with the vision to create a better living for many people.

Revenue (2023): 44.6 billion euros. Number of employees: 25.000

Product and business model:

IKEA designs and sells ready-to-assemble furniture, kitchen appliances, decoration, home accessories, and various other goods and home services, globally. Europe is the main market (70%); the USA is an increasing market (15%) as well as Asia (9%). In the last few decades IKEA has contributed to the linear economy and manufactured and marketed furniture and appliances that hold a short lifespan and limited possibility of repair and maintenance at very low prices under the philosophy that design is available for all.

Sustainability:

Vision: "As a global business with the capacity to make a large impact, we know that we have a responsibility to make that impact a positive one. At IKEA, we believe good business means ethical business."

Sustainability is anchored around three focus areas:

• Healthy and sustainable living—Inspire and enable more than 1 billion people to live a better everyday life within the boundaries of the planet.

• Circular and climate—Prioritize drastically reducing GHG emissions across the IKEA value chain and move toward the use of only renewable or recycled materials by 2030, as well as regenerating resources while growing the IKEA business.

• Fair and equal—Play our full part in contributing to a fair and equal society by respecting human rights, creating a positive impact for people across our value chain, and contributing to resilient societies.

Very ambitious communications and targets are to **only use renewable or recycled materials sourced in a responsible way by 2030**, and **a 50% reduction of absolute GHG across the entire IKEA value chain (scopes 1, 2, and 3) by 2030** and reaching **net-zero emissions by 2050**, without relying on carbon offsets to meet this absolute reduction target. The first target is ambitious but net zero in 2050 seems unambitious. IKEA will decide on the mechanism to set goals for restoration in the value chain once the Land Sector and Removals Guidance is published in FY25, thereby addressing the impacts of ecosystems and biodiversity but postponing the targets. IKEA is the largest sourcer of wood globally and documentaries are out that they are facing large challenges in their supply chain. With this position in market for wood IKEA holds a special responisbility to regenerate ecosystems and build sustainable and biodiverse forestry. This is an issue that can cost the existance of a corporation if not dealt with in a genuine way.

The communication on sustainability is explanatory and educational but in general lacks specification on metrics and targets to support the overall goals. No clear roadmap or transition plan is disclosed, and for a conglomerate dependent on production in the developing countries and to a large extend on natural ecosystems it is an extensive and open task to build the transition plan and roadmap necessary to transform and communicate a trustworthy and serious transition toward sustainability.

https://www.ikea.com/global/en/our-business/people-planet/

Gender equality in management (F/M):
EMT – 4/5 (44%); BoD – 2/7 (22%)

2.4.2 H&M Group

H&M Group
A Swedish founded corporation listed at the Swedish Nasdaq. The majority shareholder is still the family of the founder.
Revenue (2023): 21 billion euros. Number of employees: 143,000

Product and business model:
H&M Group is a global fast fashion and design company with several brands, with over 4000 stores in more than 75 markets and online sales in 60 markets. H&M's largest market is Europe, and the European and African sales are double the sales in North and South America. China and Bangladesh are the main production markets of H&M clothes. The clothes of H&M are characterized by releases, low prices, and low quality.

Sustainability:
Vision: "Our ambition is to lead the change towards a circular fashion industry with net-zero climate impact, while being a fair and equal company." They state that 85% of materials are recycled or sustainably sourced, including 25% recycled materials.
The focus areas of sustainability are:
• Leading the change—Scale innovation, promote transparency, collaborate for industrywide progress.
• Circularity and climate—become net-zero across our value chain by 2040, operating within planetary boundaries, have a net-positive impact on biodiversity, scale circular models and systems for our products, supply chains, and customer journeys
• Fair and equal—have a positive impact on people across our value chain, support and promote inclusion and diversity in everything we do
H&M has a document on their climate transition plan where they release their strategy in overview and targets to **reduce scope 1 + 2 + 3 GHG emissions by 56% by 2030**, and **90% by 2040**. H&M states that they do not use off-setting. They communicate their transition plan and disclose their value chain impacts. Targets and commitments to use: **100% recycled or sustainably sourced materials by 2030, 30% recycled materials by 2025 and 50% recycled materials by 2030, 100% polyester from recycled sources by 2025**. H&M has published corporate governance report including the materiality assessment in relation to stakeholder interests (non-double materiality assessment).
Generally, their communication is detailed, with cases, ambitions, targets, and metrics. They do not address some central issues in the fast fashion industry such as overproduction (30–50%) that never reaches the consumer and is discarded, nor the low quality of textiles and mixed fibers that challenge recycling of fibers.
https://hmgroup.com/sustainability/

Gender equality in management (F/M):
EMT – 4/3 (43%); BoD – 8/6 (43%) 4/2 employee elected representatives

2.4.3   Bestseller

Bestseller
Privately owned Danish company, owned by the family of the founders and the founders Holch
Povlsen.
https://bestseller.com/
Revenue (2023): 5 billion euros. Number of employees: 20,000+

Product and business model:
A fast fashion provider with a range of more than 20 brands, including JACK & JONES,
ONLY, and VERO MODA. They sell clothes and accessories for all ages, genders, and
occasions at low prices with many releases yearly.

Sustainability:
Vision: "Fashion FWD—'Bringing fashion forward' is our ultimate ambition that guides our
strategy, goals and actions: Climate Impact, Supporting the people in our value chain,
Preparing for a circular future."
Transition plan with targets of **reducing scope 1 + 2 GHG emission by 50%**, and **scope 3
GHG emission from purchased goods and services and upstream and downstream
transportation by 30% by 2030. No target on full scope climate neutrality.** They disclose
other targets: **100% of key waste streams from head offices and owned and operated
logistics centers will be recycled or reused by 2025.** This refers to the overproduction that
today is discarded. No targets on minimizing overproduction, but on recycling overproduction.
They disclose the climate impacts in the value chain being 1% at end of life without any
comments on the challenges in the fast fashion industry.
Very fancy communication and online interaction on sustainability but with very little facts
and addressing the large challenges of the fast fashion industry. They disclose a plan on
preparing for a circular future with various actions but **no targets on full circularity or steps
toward meeting the drafted Extended Producer Responsibility on textiles and fast
fashion.**
The targets on supporting the people in 2025 are digital payment of wages in tier 1 factories,
promoting fair living wages, and fire, electrical, and building safety. This indirectly means that
for the next 2 years this corporation will not secure their employees fair and safe conditions.
The communication on sustainability lives up to the rumors on Bestseller—a lot of fancy
window dressing and greenwashing and no ambitious targets or plans to transform into a green
and sustainable economy.
https://bestseller.com/sustainability

Gender equality in management (F/M):
EMT – 0/2 (0%); BoD – 2/3 (40%)

2.4.4   JYSK

JYSK
Danish privately owned company by the founder's family of Lars Larsen.
https://jysk.com/
Revenue (2023): 5.2 billion euros. Number of employees: 31,000

Product and business model:
JYSK is an international retailer of furniture's, textiles, interiors, and decorations for home and
garden at very cheap prices after the same linear principles as the fast fashion industry. JYSK
has over 3400 stores and operates in up to 48 countries. Corporate slogan / payoff. Always a
great offer.

Sustainability:
Vision: Seamless and Closer to the Customer. JYSK does not disclose a sustainability vision
and communicate on initiatives and the SDGs#6, 8,12,13, and 15. They communicate about
the extended responsibility that goes beyond their own employees but also employees of
suppliers. The climate target is reduction of GHG by **50% by 2030 in scope 1 + 2. No
climate-neutral targets and no scope 3 targets.** The communication on sustainability is very
poor and unprofessional and not worthy of a global corporation of this size.
https://jysk.com/sustainability/

Gender equality in management (F/M):
EMT – 1/7 (13%); BoD – 0/4 (0%)

The above companies in the fashion and lifestyle industry are significant globally
with a large market share of fast fashion and lifestyle as well as significant takers of
textiles and wood. This means that they strongly effect the state of wild nature and
biodiversity, globally as will as the supply chains in the textile, lifestyle and furniture industry. Contrary to the energy sector and the recycling industry the fast fashion and lifestyle industry is far behind in the transtion to a green and circular
economy. In the Nordics this industry has the potential to drive change in a global
perspective but are not united or focussed on the transition, maybe because the
industry is solely driven by the linear economy and take-make-waste business models and cheap, untransperant supply chains. The potential here is huge and the will
is little when reviewing their sustainability reporting and webs.

The recyclers are important in creating new value chains in the circular economy
and the transition to a circular economy is their core business. Two selected recyclers within traditional recycling are listed here:

2.4.5   Stena Recycling

Stena Recycling.
Stena Recycling is part of the Stena Sphere an international conglomerate of Stena AB, Stena
Metall (owner of Stena Recycling) and Stena Sessan. The conglomerate is privately owned by
a Swedish family, Olsson/Eriksson. https://www.stenarecycling.com
Revenue (2023/24): € 2209 million. Number of employees: 3800

Product and business model:
Stena Recycling as the significant part of the recycling business in the group is offering a
range of advanced recycling and waste management solutions, supply of recycled raw
materials for use in new products. Stena Recycling has introduced management consultancy to
catalyze the circular economy, as a new service. Stena Recycling is operating in the Nordics,
Poland, Italy, and USA with the largest activity in Denmark.

Sustainability:
Vision: "A circular partner for a more sustainable world." Stena has a strong focus on circular economy and has build their communication on facilitating the transition and also creating a responsible consumption. Recently they have published the Circular Voice 2024 to understand consumers of today. No doubt that circular economy and the transition is the business core. Stena Metall publishes an annual sustainability report, together with an annual review, on Group level. The strategic sustainability work is divided into three areas: care for the environment, care for people, and care for sustainable business. Stena Recycling publish a climate impact report and is commited to SBTi. They disclose their climate impact in scope 1+2 and 3 with 89% of the impact in scope 3. Targets are to **reduce Scope 1 & 2 with 50% by 2030**, and to **reduce absolute scope 3 GHG emissions by 25%** within the same timeframe. They have set **net-zero targets for 2050**, remaining to be validated by SBTi. They have a very significant and detailed reporting on the climate impact of SR with detailed explanations and metrics and methods. Stena Metall publish a broader Sustainability Report referred to on SR's web. This sustainability report starts addressing the transition to CSRD as the new reporting framework and is generally very linked to the new EU legislation on sustainability. This is a recommenable reporting to follow and to learn from. https://www.stenametall.com/about-us/sustainability/sustainability-reporting/

Gender equality in management (F/M):
EMT 2/7 (22%); BoD – Stena Metall 3/7 (30%)

2.4.6 Ragn-Sell Group

Ragn-Sell Group
A corporate group operating companies in Scandinavia and Estonia that is a privately and family-held corporate group.
https://www.ragnsells.com/
Revenue (2023): €528 million. Number of employees: 2900

Product and business model:
Their business areas are recycling, treatment and Detox, and new value chains, thereby being a driver of the circular transition. They collect, treat, and recycle waste and residual products from businesses, organizations, and households.

Sustainability:
Vision: "What a Waste—think circular. We want to be living proof that caring for the earth and business go hand in hand."
Their core communication and business are around the transition to a circular economy and how waste becomes a resource. They communicate sustainability around four topics:
• Value creation and innovation
• Climate and environment
• People and culture
• Responsible relations
Ragn-sell makes it clear that circular economy is the way to meet their sustainability visions and why they have an extensive and educational communication on circular economy in the center of their sustainability communication. The have targets on their main topics and they pledge to avoid **two million tons CO2e by 2030 in scope 3**. They disclose their carbon emissions including from landfills, the largest source of emission in their scope 1 + 2. They lack a clear transition plan to become climate neutral. They disclose their materiality assessment from the perspective of stakeholders, not from a business perspective as required by CSRD; still they are on good track in communicating business and ESG at the core of their business.
https://www.ragnsells.com/sustainability/

Gender equality in management (F/M):
EMT – 5/7 (42%); BoD – 1/2 (33%).

The large recyclers in the Nordics are Swedish and coming out of the metal industry and today significant in recycling. All recyclers are moving the circular economy and are important facilitators of the circular economy and often also material experts. Other selected recycling companies in the Nordics are listed here:

Company	Materials	Country	Link
NG Group	Large collector that designs, executes, and improves in every phase of manufacturing and distribution operating the Nordic, Poland, and UK	Norway	https://www.nggroup.no/
Tomra	System and technology provider of sorting and vending machines for food, waste and metal sorting system, and sorting solution for mining	Norway	https://www.tomra.com/
HJHansen Recycling Group	Recycle iron and metal waste. Extract raw materials of waste and cut-offs. They are specialized in demolition, car disassembly and recycling, wind turbine decom	Denmark	https://www.hjhansen.dk/
ZenRobotics	Provider of robots for recycling using machine learning and AI	Finland	https://www.terex.com/zenrobotics
Fortum	The recycling and waste business is part of Fortum Group—reviewed earlier in this book	Finland	https://www.fortum.com/services/recycling-waste/fortum-recycling-waste
Hellik Teigen Group	Recycling of ferrous and non-ferrous metals (iron)	Norway	https://www.hellik-teigen.no/
Hydro Volt	Battery Recycler in the Nordics	Norway	https://www.hydrovolt.com/
Dansk Retur	Recycling of drinking containers based on a deposit system	Denmark	https://danskretursystem.dk

In Chap. 3 are also included good company cases on recycling of materials into a circular economy with focus on materials difficult to recycle, e.g., mechanical recycling of plastic by AVL (Fig. 2.8).

Generally, the recycling industry is far in the circular economy and is important stakeholders and partners for the manufacturers and other stakeholders in the value chains of products.

2.5 Nordic Global Corporations

There is a big difference in the approach to sustainability and the degree to which the corporations integrate sustainability and the green transition into the business core of the corporations reviewed in this chapter. It is remarkable that only a few of the global corporates that have worked extensively with the challenges and potential of scope 3 including the circular economy, biodiversity, and ecosystems. Naturally, there is a big difference between the companies that are inherently sustainable

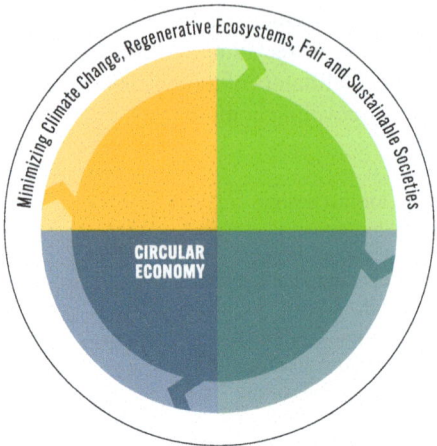

Restoring scarce resources by redesigning, reusing, and recycling.

Counteracting environmental disasters seen globally due to tremendous amounts of waste being exported from the consuming countries and deposited in the developing countries,resulting in deposits of waste as plastic, textiles, electronics with the result of pollution and other environmental disasters.

Replacing products and quality designed for the linear economy with products that are designed for the circular economy,thus increasing the potential for a sustainable lifestyle of future generations.

Controlling and replacing chemical pollutants from material streams and pollutants introduced earlier without accounting for the healthcare and environmental problems they caused.

Creating immediate energy savings from recycling compared to extraction of virgin resources.

Creating regional value chains that are transparent and traceable to ensure sustainable consumption to stabilize and develop regional economies and wealth.

The basis for regenerative ecosystems due to the decoupling of land use for mining of virgin resources.

Redesigning the global value chains to regain control of economy, and security of supply.

Fig. 2.8 Climate Nexus—Circular Economy. The Climate Nexus is a concept described in the book: The Great Transition to a Green and Circular Economy. Gitte Haar. 2024 and visualizes a holistic approach to implementing sustainability

because they deliver an environmentally friendly product, or a product that facilitates the transition on one side, and companies that must transform their business models and products on the other side. There is also a remarkable difference in how professionally companies approach sustainability and how far they go in making sustainability a business imperative and integrate it into their strategy, or just spend a lot of time and effort on communicating.

Communication Framework on Sustainability

In reviewing the corporations, it has not been important which formal framework they have used in their sustainability communication and reporting. The reviews focus on how integrated is their approach to sustainability and how thoroughly they address the real challenges, especially throughout the value chain—upstream and downstream (scope 3, according to UN Climate Gas Protocol for companies).

The SDGs are a common framework to communicate sustainability in corporations in the Nordics. Unfortunately, it seems that there often is an inverse correlation between working deeply and business-driven and the number of SDGs companies communicate about. Learn more on how companies can prioritize the SDGs in the SDG strategy house presented in Fig. 2.9. More details on the SDGs from a company as well as society perspective can be found in another book of the same author: *Rethink Economics and Business Models for Sustainability* (Haar, Rethink Economics and Business Models for Sustainability. Sustainable leadership based on the Nordic Model, 2024b).

It almost seems as if some corporations have sprinkled a stack of SDGs into the reporting because they are glamourous and have not actually worked wholeheartedly and deeply with sustainability, climate transition plans, and circular economy. It is challenging that the SDGs have received so much attention from business management without them using the SDGs strategically.

Taking responsibility for the entire value chain seems new to many of the global corporations and is still something many are struggling with. It is complex and difficult to integrate and account for the full value chain. With the new EU reporting regulation (CSRD), the large corporations must implement the extended responsibility in the full value chain and the monitoring of these due to Corporate Sustainable Due Diligence Directive (CS3D). Some of the corporations here, especially those in the energy sector, have chosen Science-Based Target Initiative (SBTi) to get their roadmaps assessed and validated. This will contribute to a greater recognition of the challenges in scope 3 and along the entire value chain but has not for all resulted in clear and ambitious targets in scope 3, yet. It is commendable that corporates are now moving away from the traditional communication approach to a science-based approach. Difficulties in understanding the full value chain are perhaps also a symptom of how silo-driven the business models of the corporates are, and how much focus there historically has been on optimizing a very small step in scope 1 + 2 rather than taking a holistic approach that sustainability requires. Nevertheless, operating in the EU now requires transition on company and product level taking full responsibility for the full value chain. This can strongly be inspired by the SDG framework and learning materials available here (https://sdgs.un.org/goals).

The European Sustainability Reporting Standard (ESRS[7]) that describes the ESG impacts of companies as the framework in meeting CSRD and the ESG sustainability standards are based on SDG as visualized in Fig. 2.10 (Haar, Rethink Economics and Business Models for Sustainability. Sustainable leadership based on the Nordic Model, 2024b).

---

[7] https://www.efrag.org/en/sustainability-reporting/esrs-workstreams

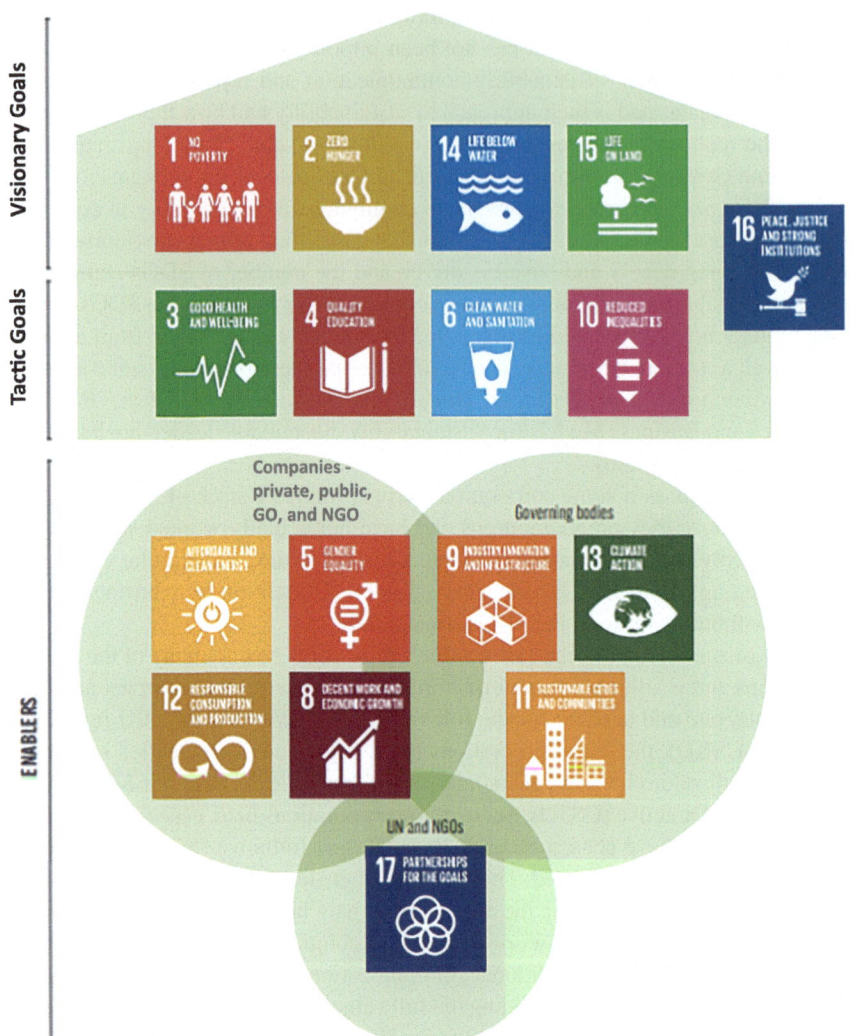

Fig. 2.9 SDG strategy house. It is important for companies to work strategically with the SDGs and only a few SDGs are relevant from a business perspective. The Strategy House helps companies and organizations prioritize

Many of the companies that have communicated on sustainability for years still use the UN Global Compact as a framework for their reporting, together with references to a very few, carefully selected SDGs. It is quite liberating that they stick to their original approach, even if they now must comply to CSRD. Almost all corporations have committed themselves to collaborations and alliances, and disclose their donations and philanthropic work, but it is still clear that the entire value chain and new business models in a circular economy mean new market conditions for companies, especially the global corporates.

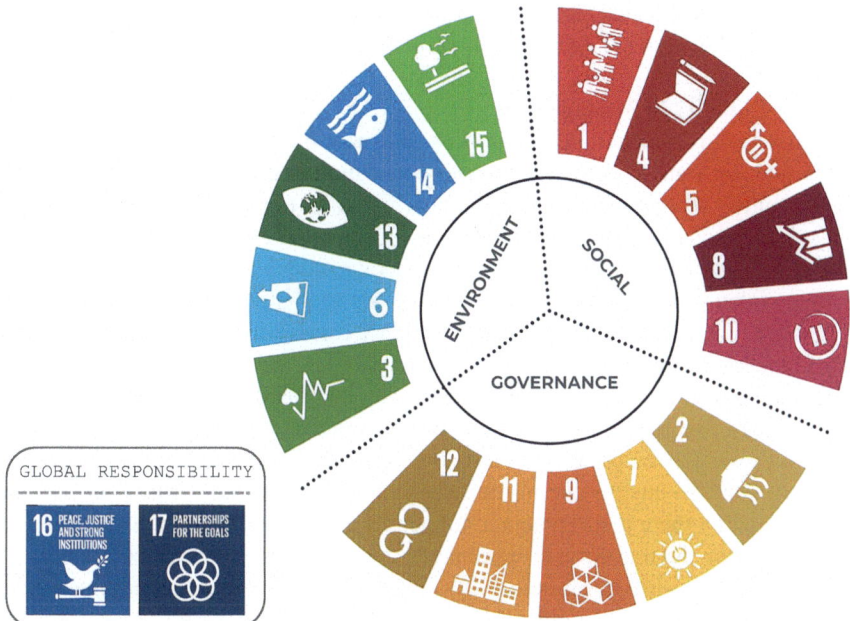

Fig. 2.10 SDG are ESG. The European Sustainability Reporting Standard (ESRS) is transforming the SDGs into a ESG framework for companies as illustrated here. The ESRS is visualized in Fig. 2.11

Foundations and Collaboratively Owned Businesses

Many of the corporations in the Nordics are owned by foundations or have established foundations where profit is directed for purpose-driven activities. Historically many of the large industry corporations in the Nordics have been family-owned until they reached a significant size. Often the companies were handed over to next generation and in this process some of the shares have been transferred to a foundation because the families wanted to give back to society from all the value they created together with employees, customers, and partners over the years. The social responsibility and thriving for a certain equality and fairness for all based on shared values, such as public healthcare, public educational systems, and public infrastructure, run in many of the companies and company owners here. Read more about sustainable leadership based on the Nordic Model in another book by the same author: *Rethink Economics and Business Models for Sustainability* (Haar, Rethink Economics and Business Models for Sustainability. Sustainable leadership based on the Nordic Model, 2024b).

There are also several collaboratives in the Nordics—corporations owned by the suppliers and/or customers as in the agricultural sector with Danish Crown, Arla, and DLG. Large retailers like COOP are also collaboratively owned by the customers. There is a tendency among younger generations to prioritize shared values, purpose-driven companies, and collaboratively owned companies when they choose workplace and establish new entrepreneurial businesses. A lot of inspiration can be found in the Nordics in respect of different kinds of ownership and purpose-driven foundations that distribute money from the profit of large corporations to the benefit of society and the planet.

Gender Equality in Management
The green transition requires innovation and new thinking throughout the value chain and with all types of stakeholders, and therefore corporates lose business opportunities when executive management contains too uniform people. This is why gender equality in management is included in the reviews of the corporates. The conclusion here is that the Nordics also lack gender equality in management as in many other countries. Despite the Nordics are seen as equal societies in respect of gender. The review here shows the Swedish corporations are in the front with most females in Executive Management Teams and Board of Directors. In Norway, regulation exists to ensure gender equality among Board of Directors with minimum 40% of one gender. This is clearly reflected in the reviews here. This is notable when reviewing 29 corporates that only two have a female CEO. Despite the expectation of gender equality due to the strong regulation in the Nordic countries on gender equality there is still potential for gender equality potentially also leading to an accelerating the transition to sustainable, fair, and purpose-driven corporates.

Denmark ranks low in various studies on gender equality in top management also in this review of Danish corporates (Mai-Britt Poulsen, 2019).

The general lack of gender equality in Nordic corporations is a challenge, because companies that do not have a gender diverse management are not able to meet a diverse group of customers, consumers, and employees, and the innovative power of companies is proven stronger when management is more diverse. It also becomes more and more clear that inclusion and gender diversity foster innovation that is so needed in the great transition to a green and circular economy. Continuously debates are on quotas to ensure gender equality of minimum 40% representation of each gender. In many other countries in and outside Europe, quotas are in place for politicians to stand for public election and it is also the case that when people are elected rather than appointed as in companies gender equality increases. There is a reason why gender quality has its own SDG#5—there is still a huge lack at all levels in society and also in management.

2.6   Conclusion on Climate Nexus in a Nordic Perspective

This chapter reviews the transition to a green and circular economy in the Nordics under the framework of the Climate Nexus and reviews 29 selected corporations in the sectors representing energy transition, transition to sustainable agriculture and healthy diets, transition to sustainable transport, and circular economy.

The Nordics are often in the spotlight to find inspiration on responsible and sustainable societies based on shared values, and the large Nordic corporations operating globally are awarded for their sustainable business approach. As seen from this Case Collection and review of the Climate Nexus in this book, there is a reason for this attention toward the Nordics. All the corporations reviewed here have taken a stake on sustainability and are communicating on sustainability and corporate responsibility to an extensive degree. Most have targets and some have transition plans to meet the Paris Agreement. Some of the corporates here are also very close to meet the reporting standards set by the EU with CSRD that they are to comply to by 2024 or 2025 dependent upon if they are public listed companies or not.

Sustainability is becoming a business imperative, and the corporations of the Nordics are embedding this into their public communication and many also into their business strategy. Nevertheless, there is still a long way to go before the business, people, and societies operate within the planetary boundaries described by the Stockholm Resilience Center.[8] The Nordics and the corporations reviewed here still face numorous challenges and have enormous tasks in the transition to a green and circular economy that can be highlighted as:

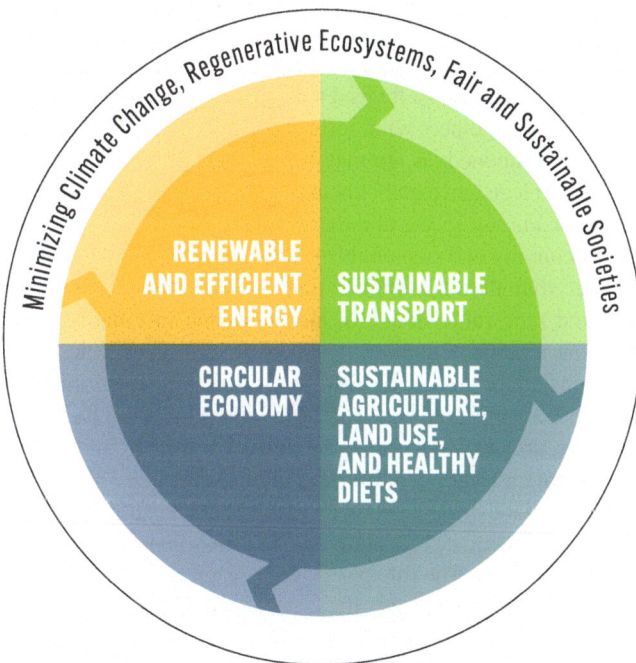

Climate Nexus

Energy transition The Nordics and the important corporations in the sector are on track in the energy transition. The region generally has low GHG emissions from energy and stands out compared to most of the world. Norway and Denmark are still extracting fossil energy sources (oil and gas) from the North Sea. Denmark has decided to stop exploitation by 2050, but Norway has not yet set an end date for extraction. Historically and for decades Finland, Sweden, Iceland, and Norway have been provided with hydrogen power and geothermal energy and have been relatively fast in the introduction of especially wind power. Due to the large forestry

[8] https://www.stockholmresilience.org/research/planetary-boundaries.html

industry in the northern part of the region, biomass has also been a part of the energy system for many years. Using biomass as an energy source is a challenge as wood and biomass is in need in many other industries in a transition towards climate neutrality, especially the construction industry. In many industries materials are substitued by biobased products and the demand for biomass is increasing dramaticly these years. Therefore the transition to climate neutrality should also include a transition away from biomass as an energy source. To some extend biowaste that does not have a potential for extraction of proteins, fibers, or lipids may be biogassed, and nutrients retained for soil fertilization. But only as the last step in a circular bioeconomy where the retention and sequencing of carbon is essential.

Denmark is in transition from coal, oil, and gas toward renewable energy, mainly wind power due to a strong industry of wind turbines, here. Now it seems as if Denmark is losing speed in the energy transition and the installation of renewable energy is not happening at the necessary speed or extent to meet the national goal of 70% reduction by 2030. The Danish Climate Counsil has just flagged that Denmark will not be able to meet this goal and the tender of a large new wind turbine park did not have any bids from operates and will now be postponed, as are many private investments in renewable energy at the moment due to legislative and bureaucratic hurdels.

The scope 3 emission of the energy sector and the dominating corporations here are still immature. Circular economy and sustainable procurement need to be implemented here at a higher speed than their communications indicate. Just because the energy delivered is clean and renewable does not mean that the installations are sustainable. Ensuring that the renewable energy installations can be recycled and materials from these large installations can be restored is still a task to onboard within the sector. Many of the raw materials used in installations, as wind turbines, solar panels, and hydro plants, are critical to drive the transition further, and therefore, it is essential to implement circular economy also in this sector.

The Hague Court of Appeals established that companies have an obligation to reduce GHG emissions and failing to take adequate reduction measures may constitute a violation of human rights and lead to liability. The ruling was based on proving the interrelation between human rights and climate change and this sentence will influence legal and court decisions around the world on the responsibilty of companies operating against the mitigation of climate change and meeting the Paris Agreement. See article on this https://www.herbertsmithfreehills.com/insights/2024-11/dutch-court-of-appeal-upholds-appeal-on-landmark-climate-litigation-case-against-shell.

Transition to sustainable agriculture, land use, and healthy diets

The transition of the agricultural and fishing industry toward a circular bioeconomy, healthy diets, and biodiversity is lacking behind the energy transition. The North Sea and parts of the Nordic land use is suffering from very intensive and industrialized production of food, and the transition plans and roadmaps to sustainability here are not clear. The national targets and strategies are poor and unaligned.

In Denmark where land areas are suffering the most from overnutrification and loss of biodiversity, there is still a blurry road, and the water ecosystems are suffering from eutrophication and overfishing. As in Norway where the ocean is suffering from industrialized production of especially salmon and general overfishing. There are no clear plans, and the corporations have no clear targets on their participation in renegerating wild nature and biodiversity. Not even if they are strongly dependent on wild nature as the food sector. Political intension has been communicated, but no strategy, implementation plans, or roadmaps are agreed upon on national levels or for the region aligned.

All the corporations reviewed here communicate on sustainability and climate targets, but the large part of the transition toward natural integrated food production is unclear, also on corporate level. Some of the corporates embed natural ecosystems and biodiversity in the communications, but clear targets, metrics, and roadmaps are lacking. Hopefully all the new frameworks and goals from the EU, Montreal, and Science Based Target on Nature will assist the corporations in integrating ecosystems and biodiversity at a strategic level.

The Finnish corporations seem to be the most ambitious in respect of nature, ecosystems, and biodiversity, probably due to ambitious political targets and strategies on this nationally.

Transparency and traceability in the forestry industry are becoming increasingly critical as the demand for wood, natural fibers, and biomass continues to rise. The urgent development and implementation of a clear circular bioeconomy, designed to sequester carbon and utilize biomass at its highest possible value, are essential. The state of our planet and the severe transgression of planetary boundaries have reached such alarming levels that human habitats and existence are now at risk. The preservation of wild nature and biodiversity is fundamental to sustaining food production, mitigating climate change, and ensuring human well-being and healthcare. Therefore, regenerating and protecting ecosystems on land, in oceans, and in freshwater systems worldwide must become an imperative for sustainable living. This goal should be enshrined in national constitutions and international laws, given its importance on par with human rights, which are already embedded in many democratic constitutions across the globe. This principle is especially relevant in the Nordic region, where some countries are beginning to take action. Aligning efforts across the Nordics is essential to drive ecosystem regeneration rooted in shared Nordic values and a deep respect for human existence and the rights of Indigenous peoples. By uniting these efforts, the Nordic region can lead by example in fostering sustainable and equitable living practices.

Transition to Sustainable Transport and Livable Cities
Sustainable transport in the Nordic Region is poor in all aspects. There are no aligned policies across the region; lack of unified investments in sustainable and shared infrastructure. Policies and commitments to minimize road transport seems like a black box from both the political side and corporate side. Some of corporates have targets on climate neutrality and commit to minimizing impacts from transport, but the roadmap to optimize road transportation and freight is poor. Even if this is the largest GHG emitter from transport, as well as a large polluter of particles and noise. It is also the transportation of people and goods on roads that makes cities

uncomfortable to live in and taking the individualized transport out of cities should free up a lot of space and fresh air. Space must be used for climate adaptation by raising forest and green areas, recreative space, and local food production.

In general there is a need for roadmaps on making cities livable and sustainable in the Nordics as all over the world. The Nordic holds some of the largest corporations within consumer brands of lifestyle, as IKEA, H&M, Jysk, and Bestseller. They are the largest takers of natural fibers and wood globally and if they united and build a sustainable supply chain and created share goals and values on regerating wild nature it would have a significant global impact on the environment and social welfare. The transition to electric vehicles is not the full solution to the large climate impacts from road transport. More shared and public transport must be available to create livable and green cities with space for recreation and urban food production. In this way, it is always important to ensure that many holistic solutions are also available in the Nordic context.

Circular economy
The Nordics are among the frontrunners in framing the need of a transition to a circular economy, compared to many other parts of the world like for example the USA. It is a strong part of the EU Green Deal as a roadmap to a strong and sustainable economy to introduce a clean and circular economy. In the Nordics, waste collection and household sorting have been implemented for years or decades, which is the first step in implementing of a circular economy. The region is, however, still producing huge amounts of waste and has a huge linear and unsustainable consumption. There is a lack of an aligned strategy for the region and Denmark does not even have a national strategy as the other Nordic countries.

Some of the large Nordic providers of fast fashion, globally have not disclosed roadmaps or specific targets for the transition to a circular economy and responsible consumption. They still hide a large blind spot of huge overproduction of low-quality cloth and lifestyle products that never meets the customer. The recycling industry is leading the transition and waste collection, and sorting is far in the Nordics, so that waste becomes available as recycled materials. Implementing closed material loops is still in the making but the necessity has been framed and the business models are slowly evolving mainly from the SME's in the region.

The next step is a detailed and clear, regional roadmap to introduce a full circular economy based on reuse of building material, clothes, packaging material, and the recycling of clean materials for next generation of recycled materials. All the corporates represented under the circular economy communicate sustainability and the issues on circular economy. The recycling industry has naturally made it their core business, but the manufacturers of products in a take-make-waste economy are lacking behind.

A recent Circular Economy Outlook (https://www.nordicinnovation.org/programs/nordic-circular-economy-outlook-2024) states that: "Business opportunities and reduction of climate emissions are two of the top drivers for circular transition among listed companies in the Nordics. A majority expect their competitors to develop significant circular capabilities. Most of these companies also believe that

EUs circular regulations will help them gain a competitive edge." This confirms that circular economy is a business imperative not only in the Nordics but in the EU already. A conclusion from the report is that more and more companies understand the power of circular strategies and are incorporating circular goals into their strategic agendas. The transition and awareness of circular economy is rising and the potential for business as well as for the environment is becoming clear. The potentials must be released and large scale implementation is still in the wakening.

Overall
Generally, Sweden seems to be the most sustainable country within the Nordics. They have for decades had a low climate impact from their energy production; they produce much less waste than the rest of the Nordics. Sweden is traditionally a country with large and strong industries, such as car, machinery, telecommunications, pharmaceuticals, and chemical goods. They also have large production of forestry, iron, and steel. Maybe the proximity to the virgin value chain and the heavy manufacturing industries have fostered more responsible consumption of energy and resources. That aligns with the view that the long and untransparent value chains across regions have driven the economy into the linear economy of take-make-waste and that shortening the value chains and having production closer to consumption will drive sustainability. It is important for consumers to understand what they purchase, and proximity is essential to drive the transition.

In the perspective of the Climate Nexus a holistic approach is necessary to transform into a fair and sustainable economy and no solutions should be evaluated 1:1 in each of the four slices. All the four slices of the Climate Nexus must be investigated when deciding on solutions. For example, the installation of renewable energy parks or farms is not necessarily sustainable just because they deliver renewable energy. The installations must be based on circular business models that ensure reuse and recycling of the installations and materials infinitely, and in respect of wild nature and local communities.

Farms and parks often take up land in regions where wild nature is scarce; this is not sustainable. We must leave land available for wild nature and biodiversity to flourish. Especially since many buildings, roofs and guard rails along roads and tracks are available to place the turbines and solar panels on in an integrated manner. Another example of the holitic approach of the Climate Nexus is that a sustainable diet with low environmental and climate impacts is often also healthy diets. So when transforming the Nordic land use, agricultur, forestry and fishery it is important with an integrated and holistic approach that creates a fair and sustainable living broadly.

The Climate Nexus reflects the EU Sustainability Reporting Standards (ESRS) in Fig. 2.11, which companies operating in the EU, and in the Nordics, are subject to. The European Sustainability Reporting Standards (ESRS) are based on the Sustainable Development Goals and are becoming a global standard for sustainability, and thereby the frameworks for sustainability in companies become more and more aligned to address sustainability and ESG. It seems as if the European

Sustainability

Cross-cutting standards		Environment		Social		Governance	
ESRS 1	General requirements	ESRS E1	Climate change	ESRS S1	Own workforce	ESRS G1	Business conduct
ESRS 2	General disclosures	ESRS E2	Pollution	ESRS S2	Workers in the value chain		
		ESRS E3	Water and marine resources	ESRS S3	Affected communities		
		ESRS E4	Biodiversity and ecosystems	ESRS S4	Consumers and end-users		
		ESRS E5	Resource use and circular economy				

Fig. 2.11 EU Sustainability Standards (ESRS). The ESRS are based on the Sustainable Development Goals and are becoming a global sustainability standard and are the framework for reporting requirements (CSRD and SFRD) and the Sustainable Product Regulation in the EU (SPR)

Sustainability Standards are becoming a global standard as corporations overseas in Northern America and Asia are looking in this direction when demanding alignment within definitions of sustainability.

Sustainable Development Goals (https://sdgs.un.org/goals)

References

Finland, M. o. (2024). https://www.treasuryfinland.fi/. (State Treasury Republic of Finland) Hentet fra Carbon Neutral Finland 2025: https://www.treasuryfinland.fi/investor-relations/sustainability-and-finnish-government-bonds/carbon-neutral-finland-2035/

Haar, G. (2024a). The great transition to a green and circular economy. *Climate nexus and sustainability*. SpringerNature.

Haar, G. (2024b). *Rethink economics and business models for sustainability*. Springer.

Mai-Britt Poulsen, M. P.-L. (2019). *Wake up*. BCG.

Case Collection of Innovative Small and Medium-Sized Circular Enterprises

3

3.1 Small and Medium-Sized Enterprises Show the Way

The innovative, small and medium-sized enterprises (SME) in this chapter of the Case Collection represent different sectors and industries. Companies from the building materials industry, the recycling industry, the textile industry, and the packaging industry are represented here. Two corporate cases has found their way into this chapter because they are important to showcase. Now, the construction industry and the packaging industry are some of the industries that meet legal requirements for green and circular transition and the first industries to transform due to the EU Green Deal and the Circular Economy Action Plan (CEAP). These two industries generate a lot of waste, and a transformation will mitigate some of the major negative impacts on the planet in terms of climate change, resource efficiency, and on wild nature and biodiversity. This Chapter includes 10 company cases for inspiration, especially into the circular economy.

The cases here are mainly on circular economy and showcase the SMEs as frontrunners in the transition. Large corporations are maybe more cautious in marketing the initiatives on circular economy as they risk cannibalizing their existing business, so they will often delay the introduction at full scale and test the circular solutions in more safe spaces. To look for inspiration, several good cases on strategic development of new business models as well as products are necessary and hopefully the cases here may do so.

Two corporations are included in this chapter. Lindstrôm, as they are implementing a circular business model in the textile industry that is important to showcase. This is very important to counter the lack of transition in the fast fashion and lifestyle corporations presented in Chap. 2. The other corporation Plus Pack is working politically to change the design and value streams of the plastic food packaging industry that is very challenged and were one company cannot solve the transition but the work needs to be done in collaboration with the industry and the value chain. Meldgaard Greenline and KLS are SMEs but part of a group not defined as SME.

© The Author(s), under exclusive license to Springer Nature
Switzerland AG 2025
G. Haar, *Nordic Case Collection on Sustainability and Transition to a Circular Economy*, Springer Business Cases,
https://doi.org/10.1007/978-3-031-78638-9_3

All the industries—corporates as well as SMEs in this Case Collection—are characterized by long value chains and a high degree of imported raw materials to the region. Therefore, these industries have a major task in both working upstream with their supply of raw material and downstream. Most industries risk disruption from new and unknown players if they do not start the transition now. All industries face challenges they cannot solve alone. Collective and collaborative efforts must be made along the entire value chain including a wide range of stakeholders, such as customers, end users, suppliers, municipalities and states, and financial partners to create sustainability and circular material loops based on traceable and transparent supply chains.

Companies from all industries need to be inspired to develop new business models and new business cases to get started with the green and circular transition. More on this can be found in another book by the same author: *The Great Transition to a Green and Circular Economy*—a handbook on transforming arose from the work with some of the companies in this chapter for the Case Collection (Haar, The Great Transition to a Green and Circular Economy. Climate Nexus and Sustainability, 2024). The companies represented here often talk openly about the challenges and barriers of the transition to a green and circular economy. It is important to listen to these companies and the challenges they met, and what they point at as structural barriers for the green and circular transition. Challenges and barriers that are only solved in collaboration along value chains, across sectors, with trade associations and municipalities. And by politicians daring to create long-term conditions for companies seriously to invest in the green transition.

The global corporations are good at communicating sustainability and ESG. They excellently publish impressive annual reports including sustainability. They are good at working extensively and systematically with sustainability and the communication of it. But generally they are not as far with the transition to circular economy and to embedding ecosystems, wild nature, and biodiversity into the core of their businesses and products as the SMEs. Some of the global corporates are an intrinsic part of the green transition, such as Novonesis, a merger of Chr. Hansen and Novozymes (https://www.novonesis.com/en). Another large corporation in the Nordics that are known for their responsible and sustainable approach towards children is Lego (https://www.lego.com/). Lego has put a lot of money behind its transition to circular economy looking for biodegradable raw materials for their plastic bricks. This might be a dead end as the demand for biomaterials is rapidly increasing, and Lego bricks are made from a good, long-lasting, and recyclable material already. Lego should rather pursue a circular business model that ensures that the used bricks are returned, and circular business models are created to ensure reuse and recycling of the materials. If Lego chooses to go with bio-based material composting at the end of life must be ensured otherwise they still have to create circular material loops to meet the new product regulation (ESPR). Nevertheless, Lego has an impressive communication on responsibility and sustainability and also has a foundation that supports learning through play very much in line with the name: *LEg GOdt*, meaning play good in Danish (https://learningthroughplay.com/).

Some of these Nordic global corporations obviously have potential new circular business models with great financial potential in a green and circular economy that have not yet been implemented, nor communicated about. This is also proven by the new Circular Economy Outlook where a large number of listed corporations evaluate the market potential of a circular economy (https://www.nordicinnovation.org/programs/nordic-circular-economy-outlook-2024).

The cases in this chapter are characterized by their tenacity and thorough work with their products and business strategies, and not least with the documentation of the impacts of the new circular products rather than the sustainability reports of the company. The SMEs here are in stark contrast to the greenwashing that customers and the market unfortunately encounter a lot these days.

The overall conclusions from the cases here are that if the business idea, products and business model is genuine, convincing and a good business case for the future, then external reporting on ESG and sustainability becomes less significant. Most of the SMEs here have sustainability in the core and have not a wide range of commitments, sustainability framework or documentation. It is important still to support, cheerish and promote these companies because their solutions are important in the transition. Most important is it to procure their products. The companies mention the main barrieres as: that lack of political will, lack of strong market legislation, and lack of competences and understanding with the customers. Some also mention that low prices of the traditional, linear products as a barriere. Pricing the negative impacts and the externalities is important to accelerate the transition. Building market conditions based on green and circular solutions is a big challange and if these SME's do not find a market and costumers that dare invest in innovative partnerships with then these good solutions will not scale. Good learnings and recommendations are available in the cases provided here.

The companies included in this Case Collection are:

Company	Industry	Sustainability category	Country
Lindström	Textile rental Service	Corporate Circular Business in the textile industry	Finland
Circularity	Collection and recycling of textile fibers	SME Circular Business in the textile industry	Sweden
Natural Indigo	Manufacturing of natural based dye for textiles	SME Circular Business in the textile industry	Finland
Meldgaard Greenline	Recycling of sand from used glass	SME part of Corporate Circular Business in the recycling industry	Denmark
Plus Pack	Manufacturer of food packaging from plastic and aluminum	Corporate in the plastic industry	Denmark
AVL	Mechanical recycling of plastic	SME Circular Business in the plastic industry	Denmark
KALK	Manufacturer of recycling of mortar and paint from lime	SME Circular Business in the construction industry	Denmark
FabricAir	Manufacturer of textile ducts for ventilation	SME Circular Business in the construction industry	Denmark
Valtavalo	Manufacturing of replaceable LED light pulps	SME Circular Business in the housing industry	Finland

Company	Industry	Sustainability category	Country
OnePan	Household product—recycling of pans	SME Circular Business in the consumer and household products industry	Sweden
KLS Pureprint A/S	Manufacturing fiber-based packaging materials	SME Circular Business in the packaging industry	Denmark

3.2   Lindström

Lindström Group Hermannin rantatie 8 00580 Helsinki Finland https://lindstromgroup.com	

Products, services, and business model:	**Framework of sustainability:**

Products, services, and business model:
Textile rental service. We design, manufacture, maintain, and wash rental textiles that we deliver to our customers. Our entire business model is based on circular economy principles

Value Proposition in short:
The customers can focus on their core business while we take care of their textile needs with our easy-to-use and sustainable services—ranging from workwear and cleanroom textiles, mats, industrial wipers, and washroom products to textiles for hotels, restaurants, and healthcare. Our comprehensive service is an effortless and care-free solution for our customers. It includes textiles, their washing and maintenance, and recycling at the end of their lives
Sustainability is at the heart of who we are. It is in the interest of our business to avoid overproduction and optimize the use of natural resources. All our decisions are guided by our purpose of caring for people and our planet by inspiring people to shine and businesses to grow in a sustainable way We always strive to raise the bar to ensure the most environmentally friendly services for our customers now and in the future

Framework of sustainability:
SDGs, SBTi, GRI, CSRD, CDP
Selected impact areas:
Circular economy, climate change, water and marine resources, own workforce, pollution

Short **description of the positive impact that your sustainability work has created on the planet and the people by products, deliveries, and the transformation toward a sustainable business:**
Circular economy is embedded in our business model. Our textile services reduce the need for new textiles while offering quality, comfort, and safety for its users. We conserve natural resources by optimizing the use of water and energy in our laundries and by optimizing our customer deliveries. Lower washing temperatures, shorter process times, and efficient energy use have improved energy efficiency in our laundries in the past few years. We utilize the heat from wastewater and use it for heating freshwater in the washing process. And when textiles reach end of their life, we recycle them as new products, as raw material for different industries, or as fibers back to the textile industry. By 2025, we aim to recycle 100% of our textile waste. In 2023, our net-zero targets were approved by the SBTi, and we are on the way to reduce our GHG emissions to reach net-zero

Sustainability certificates and ecolabels that the company uses:
Ecovadis, ISO 9001 for Quality, ISO 14001 for Environment, ISO 45001 for Health and Safety, EN 14065 Hygiene and biocontamination control. Öko-tex certificate is a basic requirement for the textiles we use. The Washroom Services in Finland operate in compliance with Nordic Swan Ecolabel

Sustainable Value Proposition/Customer Impacts: Our customers do not need to buy textiles themselves and often renting textiles (e.g., workwear) results in fewer textiles needed Textiles are also repaired and washed in a proper manner, which extends the lifespan of the products. Our washing process also optimized savings of water, energy, and detergent	**Sustainability Category** Circular business model and SDG Corporate
Revenue (2022/2023 full year) (€): 501 million **Number of Employees (2022/2023):** 5055	**SDG #12. Circular business model:** We aim to leave zero burden to the environment by reducing overproduction and conserving natural resources. We reduce overproduction by extending the lifetime of textiles and by manufacturing additional orders on demand in our own manufacturing facilities, Prodems. By using durable materials and designing for circularity we promote sustainable practices at the very beginning. Furthermore, by repairing and reusing our garments we avoid millions of kilos of unnecessary textile production We conserve natural resources by optimizing the use of water and energy in our laundries and by optimizing our customer deliveries. Lower washing temperatures, shorter process times, and efficient energy use have improved energy-efficiency in our laundries in the past few years. We utilize the heat from wastewater and use it for heating freshwater in the washing process. And when textiles reach end of their life, we recycle them as new products, as raw material for different industries, or as fibers back to the textile industry. By 2025, we aim to recycle 100% of our textile waste
Ready for the Green and Circular Economy since: Lindström is founded in 1848, but started the textile rental services in the 1930s, which could be regarded as the beginning of our circular journey. The business model helps our customers reduce textile waste, as they won't need to purchase textiles themselves, but get them as a service instead	**Sustainability take-off position:** It has been in our business interest to act in a sustainable manner, keeping textiles long in use and optimize the use of resources. So, we have done these kinds of actions for a long time, like using the poly-cotton blend fabric for its better durability in 1950s In 1991, Lindström became one of the first companies in Finland to sign the ICC Business Charter for Sustainable Development, marking the initiation of systematic environmental management. Our first ESG report was published for the reporting year 2002

Biggest achievement with sustainability in the marketplace:	**Significant impact measures (non-financial/ESG) achievements on environmental, social, and governance impacts:**
In 2023, we recycled 74% of our textiles and we aim for 100% in 2025 In 2023, our targets to halve our greenhouse gas emissions across the value chain by 2030 and reach net-zero emissions by 2050 approved by the Science Based Targets initiative. This marks the beginning of an ambitious journey to reduce our GHG emissions across the value chain while growing our business in Asia and Europe We have reached EcoVadis gold rating twice (2022, 2023)	The group target for textile recycling rate is in 2024 to recycle 90% of the textiles and in 2025 100%. The progress made here is followed to meet these targets both on a group and on a region level, but also per service line and per laundry. We also measure the amount of recycled textiles used in new products, which we aim to get to 30% by 2025 Water usage is measured per service line and the aim is to optimize water usage by utilizing the heat from wastewater and use it for heating freshwater in the washing process. Lindström have managed to reduce the water and energy usage per washed textile kilogram by 50% in the last three decades Our GHG reduction targets have been validated by SBTi and we are measuring our progress annually
Which stakeholder(s) are the drivers of the transition to sustainability in the company: Management, board, customers, employees	**Which stakeholders are the most affected by the transition to sustainability:** Employees, suppliers, management

What were the largest barriers in introducing a new business model/product in the company and in the marketplace:
Some barriers include educating the market of the new business model and monetizing its value. On many of our markets, the textile rental model is new. That requires a change of mindset from our customers, moving from the culture of owning to the culture of renting

Author's comment (Gitte Haar):
Lindström is an example of a global corporation that is driving toward circularity at the core of their business and that has seen the necessity to transform due to market potential but probably also due to the newcoming legislation (Extended Producer Responsibility + ESPR) on textiles. They may get the benefit of a first-moving position. Lindström holds a large part of the value chain in their own control, and this gives them the position to transform and include customers and suppliers in the journey. We must see more corporations in the large industries act and transform into circularity, which is why this case is important here. Many corporations are preparing for the green and circular transition, but the launch of new products and services is often tested in limited markets or with few customers before being marketed widely. As the green and circular solutions might cannibalize the existing business of these corporates.

3.3 Circularity

Circularity Sweden Traneråsvägen 7 374 95 Trensum Sweden https://circularity-works.com/	Circularity
Products, services, and business model: Circularity Sweden provides the collection of used textiles from customers; the processing and recycling of the collected fibers; the production and sale of new garments made exclusively using recycled fiber **Value Proposition in short:** Circularly produced 100% recycled clothing	**Framework of sustainability Sustainable Development Goals:** ESG and soon also B-corp **Selected impact areas:** SDG#8, 12, and 13

Short description of the positive impact that your sustainability work has created on the planet and the people by products, deliveries, and the transformation toward a sustainable business:

Just one recycled t-shirt represents a saving of 2943 L of water, 0.17 kg of cotton, 0.17 L of oil, 0.43 kg of chemicals, and 1.51 kg CO_2 emissions. And each recycled t-shirt can itself be recycled up to five times. Given forecast sales volumes of around 1 mio. pieces annually, the potential is phenomenal

Sustainability certificates and ecolabels that the company uses:
ESG, GRS, soon also B-corp

Sustainable Value Proposition/ Customer Impacts: Greatly reduced use of resources, reduction of waste through reuse	**Sustainability Category:** Circular business model and SDG startup
Revenue (2022/2023 full year) (€): Circularity Sweden came into existence in late 2023 **Number of Employees (2022/2023):** 2	**SDG #12. Circular business model**: Corporate clients collect textiles (uniforms, etc.) that can no longer be used and send them to Circularity, where they are color-sorted, stripped, and mutated before being sent to our factory for shredding The resulting fiber is spun into yarn without the addition of any virgin fiber whatsoever, from which fabric is knitted or woven. We then produce new garments for our customers. The process requires no water or chemicals
Ready for the Green and Circular Economy since: 2021 (in the Netherlands); 2023 (Sweden)	**Sustainability Take-off position**: Decades of experiencing first-hand the pollution and unfair/unsafe conditions in the textile industry
Biggest achievement with sustainability in the marketplace: Greatly reduced use of resources; provision of fair and safe working conditions; the opportunity to reuse fibers up to five times	**Significant impact measures:** Focus on true recycling under fair and safe conditions; zero use of water, chemicals, and virgin fiber. We do not compromise to get more orders!

Which stakeholder(s) are the drivers of the transition to sustainability in the company:	Which stakeholders are the most affected by the transition to sustainability:
Circularity is inherently sustainable; there is thus no transition to sustainability—we simply *are* sustainable	Our employees who carry out the recycling and production of new garments are positively affected by fair and safe working conditions; our customers benefit from improvements to their environmental footprint and true compliance with ESG regulations. We issue certificates that can be used for environmental audits; the environment benefits from greatly reduced use of resources

What were the largest barriers in introducing a new business model/product in the company and in the marketplace:
The greatest hurdle is convincing clients of the need to switch from and the benefits of a circular as opposed to a linear system

Other comments:
Circularity Sweden was only established in late 2023, so aspects such as the collection of old textiles from clients are still under development. The goal is nonetheless clear, the aim being to carry out as many steps of collection, recycling, and production as close as possible to the point of use of the textiles.

Author's comment (Gitte Haar):
Circularity is one of many examples of a startup newly established in Sweden that wants to recycle fibers for clothes into new recycled fabrics. There is a great need for a lot of new initiatives to solve the lack of circularity in the textile and fast fashion industry. Circularity is based in the Netherlands and is now also becoming established in the Nordics

3.4   Natural Indigo

Natural Indigo Finland Järvikyläntie 850 85560 Nivala-Finland www.naturalindigo.fi	

Products, services, and business model:	Framework of sustainability.
Natural Indigo Finland produces natural dyes from industrial by-products, such as waste coffee grounds from coffee roasteries. We offer the opportunity to replace oil-based colors with natural dyes. Applications include the textile and fashion industry, biomaterials, packaging materials, and various ecological coatings **Value Proposition in short:** We generate tangible sustainable development by providing natural dyes from industrial by-products, thereby replacing the use of synthetic colors, and reducing carbon dioxide emissions and pollution caused by synthetic dyes in water bodies	Our company's operations are based on principles of sustainable development, particularly focusing on utilizing industrial by-products to reduce environmental impacts

Selected impact areas:
Environmental Impacts: We reduce the use of synthetic colors and carbon dioxide emissions
Social Impacts: We enable the practical implementation of sustainable development in business operations
Economic Impacts: We create new business opportunities by utilizing natural dyes

Short description of the positive impact that your sustainability work has created on the planet and the people by products, deliveries, and the transformation toward a sustainable business:
Our production reduces environmental harm and offers our customers more sustainable alternatives. Additionally, by utilizing industrial by-products, we promote the circular economy. The problems with synthetic colors are that they carry significant GHG emissions, environmentally harmful chemicals, water pollution, contamination of drinking water and water systems. Natural colors will mitigate all these negative environmental impacts.

Sustainability certificates and ecolabels that the company uses (both product labels and company certificates):
n/a

Sustainable Value Proposition/Customer Impacts: Our customers can contribute to the green transition by choosing natural dyes, and their experiences are positive. Simultaneously, we utilize industrial by-products, increasing customer satisfaction with sustainable choices. The natural colors are suitable for a lot of industries, as: textile and fashion, biomaterials, packaging, ecological coating.	**Sustainability Category:** Circular business model and SDG startup with market traction
Revenue (2022/2023 full year) (€): 2022: €20,000 2023: €80,000 **Number of Employees (2022/2023):** 2022: 1 / 2023: 1	**SDG #12. Circular business model:** NI is a facilitator of a clean and circular economy relaying on wild nature.
Ready for the Green and Circular Economy since: 2019	**Take-off position:** Sustainability is the business core in Natural Indigo and has been since the start of the company.
Biggest achievement with sustainability in the marketplace: Commercial collaboration with significant brands such as Marimekko and Lapuan Kankurit. Additionally, we collaborate with coffee roastery Paulig, strengthening the perspective of sustainable development and creating new business opportunities	**Significant impact measures (non-financial/ESG) achievements on environmental, social, and governance impacts:** n/a
Which stakeholder(s) are the drivers of the transition to sustainability in the company (management, project managers, etc): All employees, suppliers and customers	**Which stakeholders are the most affected by the transition to sustainability (management, customers, employees, suppliers, etc.):** All stakeholders as it is the core of the business

What were the largest barriers in introducing a new business model/product in the company and in the marketplace:
Lack of competences, lack of data, lack of involvement internally/externally, lack of take-back system, etc.

Author's comment (Gitte Haar):
This case is important because all steps in the value chain of the textile industry and other industries needs to be transformed and prepared for a clean and circular economy. Providing natural dyes produced responsiblely and in respect of nature is important since many industries struggle with hazardous chemicals from mineral oils and additives. Natural colors counter many of the healtcare and ecosystems issues caused by synthetic colors. This can also be seen from the partners / consumer brands that has teamed up with NI. Natural Indigo has a large potential in unfolding new markets and products. This case of a small startup confirms that being born into a green and circular economy does not leave a lot of time and energy for storytelling or ESG reporting, which is not necessary when the products and business model is clear and obvious for customers.

3.5   Meldgaard Greenline

Meldgaard Greenline A/S Askelund 10 6200 Aabenraa	

Products, services, and business model:

Meldgaard Greenline (MG) operates several types of businesses within recycling of blasting agents for sand cutting processes, recycling from glass and residues from coal-fired power plants, other abrasives (many recycled), blasting machines/cabinets, spare parts, and other equipment typically B2B.

Blasting and water cutting agents from silicate

For 40 years, MG has sold blasting agents either silicate from the coal-fired CHP plants or extracted virgin minerals from mines outside Europe. Both products are now a scarce resource. The industry therefore urgently needs alternative high-quality products on the market. Customers who purchase cutting sand for waterjet cutting are offered to return the used cutting sand. The sand is cleaned and recycled to new cutting sand or as a new blasting agent for the steel and marine industries, as well as abrasives for waterjet cutting.

Compared to both the "Alusilicate" from coal and the "Garnet" cutting sand, landfill is the only alternative for the used sand unless recycled by MG.

Residues from coal-fired power plants in Europe are sold as blasting agents to the steel industry in the Nordic countries. The used blasting agents are then taken back, purified, and incorporated as a raw material for insulation material.

Blasting agents from recycled glass

The glass that is used is the "worst" glass; i.e., it cannot be recycled as glass at the smelters. Much of this glass is currently transported abroad, or some municipalities even allow the glass to be buried and deposited uncleaned in the ground (landfill). Neither is a good solution for the environment. The glass that is converted into blasting agents comes from AFLD (a joint municipal partnership). The glass has been tested and has exactly the hardness and edges of a good blowing product. At the same time, it is not harmful to health, unlike the previous quartz sand that has also been used in the industry, and which is not allowed in the EU as there is a risk of silicosis and cancer.

MG receives waste glass from businesses, local authorities, and waste management companies. The waste glass is contaminated with ceramics and for this reason cannot be recycled at glassworks, such as many wine bottles and jam jars. The glass is crushed, sorted, washed, and cleaned, at 450 °C. The high temperatures burn off all organic material that is left over in the crushed glass. Once the crushed glass has been processed, it is sieved into various grainsizes and sold as blasting abrasive. This blasting abrasive is a good alternative to traditional blasting grits, which are often based on scarce resources sourced from outside of Europe, and it also addresses any potential challenges with workplace safety, as it is not harmful to one's health. Glass is a 100% recycled product, which at the same time triggers challenges with the working environment. Meldgaard Greenline (MG) built a factory where old wine bottles and jam jars are crushed and burned to make it a high-quality material for sandblasting.

Value Proposition in short:

Environmentally friendly blowing and cutting agents.

Meldgaard Greenline A/S is a leader in Denmark within blasting agents for the steel and marine industries as well as abrasive for waterjet cutting.

MG offers full solutions of transport, take-back of used goods, and sale of spare parts for waterjet cutting machines.

Framework of sustainability:
Just started using ESG for the entire organization, so not a lot of data for Meldgaard Greenline yet.

Selected impact areas:
In regard to Meldgaard Greenline's glass factory—it receives a steady stream of waste glass due to the recent waste program the Danish government has implemented—this means that companies and citizens must sort their waste into certain fractions, one of them being glass. The finished product ReBlast is more interesting to many customers as it is a 100% recycled product, which to many customers is an important factor.

***Short* description of the positive impact that your sustainability work has created on the planet and the people by products, deliveries, and the transformation toward a sustainable business:**
The glass factory and its products offer a new sales channel to companies that collect waste glass—but it also creates a new 100% recycled product that meets the requirements from customers asking for efficient and quality products that are better for the environment (locally made, recycled, not harmful to the environment) compared to many alternatives.

Sustainability certificates and ecolabels that the company uses:
None

Sustainable Value Proposition/customer impacts: The glass factory offers its customer a 100% recycled product that is efficient and safe to use. On some abrasives we offer a return scheme where we take back the used abrasives to ensure it is handled correctly as a waste product. We also offer abrasives that can be used, multiple times	
Revenue (2022/2023 full year) (€): Entire Meldgaard Group: 2022: 122 mill € 2023: 128 mill € **Number of Employees (2022/2023):** Entire Meldgaard Holding approx.: 770	**SDG #12. Circular business model:** The core of MG's business is all circular business models and one of the products, SuperCut, is purified garnet sand, whose cutting performance improves through recycling. This involves "upcycling" of the resource. As a supplier, we help the industry reduce the need for extraction in mines by reusing the products. We offer a complete solution that optimizes customers' production processes while ensuring environmentally correct disposal of waste
Ready for the Green and Circular Economy since: Meldgaard Greenline has offered products the customer can return after use since 2006. The glass factory started producing ReBlast in 2021	**Take-off position:** Meldgaard has always strived to find solutions that make sense from both a financial standpoint and an environmental perspective
Biggest achievement with sustainability in the marketplace: Finding a use for discarded glass and turning it into a new and efficient product that customers want because of its quality, price, and it being a recycled product	**Significant impact measures:** N/A we have no data on this

Which stakeholder(s) are the drivers of the transition to sustainability in the company: Upper management and head of departments and employees with daily customer contact	Which stakeholders are the most affected by the transition to sustainability: Employees, suppliers, and customers

What were the largest barriers in introducing a new business model/product in the company and in the marketplace:
Lack of data—lots of testing and adjusting were necessary.

Other comments:

Author Comments (Gitte Haar):
Greenline is part of Meldgaard Group that is a very innovative group within recycling and development of new technologies for recycling and upcycling of resources from waste. Greenline with the business in recycling of sand from that fraction of glass that cannot be reused/recycled as glass proves that looking at waste as a resource, large profitable potential for recovery evolves. Meldgaard Group and Greenline are characterized by their ability to identify and develop scalable and profitable solutions. Sustainability here is a perquisite in all they do. Meldgaard Group is an example of the genuine sustainability in the span field of technology and business. Meldgaard is driven by business and see sustainability as a business opportunity, and modest responsibility is the core of their business conduct. This is not a group in transition to a green and circular economy. This is a group that is born sustainable and where sustainability is at the core of business. These types of companies are often not noticed and awared for the great work they do. Their contribution to the green transition is common business conduct rather than communication and annual reports. These companies should to a much larger extent be promoted, cheered, and acknowledged for the great work they do and we need these cases to illustrate and inspire others.

3.6   Plus Pack

Plus Pack AS *Energivej 40* *5260 Odense S* www.pluspack. com	

Products, services, and business model:
Plus Pack designs, develops, manufactures, and sells packaging solutions for food, both branded solutions and customized solutions. Key business areas are aluminum and plastic packaging
Key applications are frozen, cold, and warm food and ready meals. Key segments are industry, food service, horeca, and retail.

Value Proposition in short:
Plus Pack (PP) enables customers to protect food and materials, using fewer of our planet's resources by continuously developing food packaging solutions that are easy to recycle or reuse after use
Plus Pack work with customers, suppliers, end users, and other relevant stakeholders to identify packaging solutions or processes, which can help reduce the environmental impact

Framework of sustainability:

Plus Pack focuses on SDG#12 (Responsible consumption and production) and four selected subtargets, which give strategic direction to the company's long-term sustainability efforts: 12.2 (We minimize footprint); 12.3 (We fight food waste); 12.5 (We think circular); 12.8 (We engage and inspire)

Since 2021, Plus Pack has voluntarily integrated core sustainability data into the management's review of its annual report, covering strategic company KPIs within ESG, e.g., CO_2 emissions, energy consumption, and recyclable materials as well as sick leave, accidents, employee well-being scores, and diversity.

Plus Pack uses the international standard the GHG Protocol to classify and calculate its climate emissions

With the approval of Plus Pack's near-term science-based emissions target by the Science Based Target Initiative in 2022, Plus Pack is committed to location-based CO_2 emission reporting

Plus Pack uses ASI-certified aluminum suppliers

Selected impact areas:

Plus Pack's Double Materiality Assessment has identified and prioritized the most sustainable impacts, risks, and opportunities with a special focus on the value chains of aluminum and plastics and product safety in terms of consumer behavior

***Short* description of the positive impact that your sustainability work has created on the planet and the people by products, deliveries, and the transformation toward a sustainable business:**

Plus Pack is engaging in public–private partnerships and actively participating in collaborations and projects, which aim at substantially reducing packaging waste generation through an increased focus on the prevention, reduction, reuse, and recycling of food packaging materials—locally, nationally, and internationally.

In 2023, Plus Pack reduced its absolute CO_2 emissions by 60% compared to baseline year 2019 (scope 1+2) and increased its use of production materials which are easy to recycle to 99.7% (scope 3).

On top, Plus Pack invests time and resources in sharing knowledge, experience, and insights with a variety of stakeholders through offering our own Packaging School sessions, chairing one of the Danish Government's 14 Climate Partnerships and participating in the national Green Business Forum, co-chairing the Retail Sector Cooperation on Plastic Packaging under the Danish Ministry of Environment

Sustainability certificates and ecolabels that the company uses:

Plus Pack is certified:

ISO 14001:2015 and 9001:2011

BRC Global Standard for Packaging materials and Normpack

In 2022, Plus Pack invested in a new warehouse building, which was certified Silver according to the standard by the German Sustainable Building Council (DGNB). Also, in 2022 Plus Pack had its near-term science-based CO_2 emission target approved by the Science Based Targets Initiative.

In 2024, Plus Pack has become a Downstream Supporter member of Aluminium Stewardship Initiative (ASI) in 2024

Sustainable Value Proposition/customer impacts:

Plus Pack enables customers to protect food and materials, using fewer of our planet's resources by continuously developing food packaging solutions that are easy to recycle or reuse after use

Revenue (2022/2023 full year) (€): Net revenue (2023): t€ 87.785 **Number of Employees (2022/2023):** Average number of employees (2023): 191	
	SDG #12. Circular business model: *Plus Pack believes that food packaging is part of the long-term solution. We work to prevent waste in all activities, minimize footprint, and increase resource productivity while reducing overall carbon emissions and growing a product assortment in materials, which are easy to recycle in accordance with the guiding principles for material recycling within a circular economy.* In the annual report 2023, we state how we move toward more circular products and reduce CO_2: 1. Push for circular products. Examples from 2023: completed 52 projects with customers to move toward more circular products. Held 29 packaging school sessions. Introduced plastic products for reuse 2. Design for recyclability. Examples from 2023: designed new products in clear PP and clear recycled PET 3. Reduce CO_2 at our sites. Examples from 2023: reduced scope 1 and 2 emissions by 29%. Reduced CO_2 emissions per kilo material used in production by 9%. Decided not to make use of available CO_2 certificates for CO_2 reduction 4. Reduce CO_2 in the value chain. Examples from 2023. Calculated scope 3 baseline. Reported scope 3 emissions in Annual Report. Selected low carbon aluminum sourcing as key initiative for 2024
Ready for the Green and Circular Economy since: 2019	**What initiated the strong focus on sustainability, ESG, or SDG:** The launch of the 17 SDGs in September 2015 and the fourth-generation owners taking over the company gave the needed platform to dedicate more focus on long-term sustainability change both in business and in society (systemic)

Biggest achievement with sustainability in the marketplace: Documentation of recycled content in materials, reporting on CO_2 emissions with Science Based Targets Initiative and Circular Packaging School sessions with customers	**Significant impact measures:** Since 2021, Plus Pack has used the free evaluation tool Circulytics provided by Ellen Macarthur Foundation and scored a "B" in 2023, which was above industry standard and the best result in 3 years. From the annual report 2023:

SUSTAINABILITY HIGHLIGHTS FOR THE GROUP

KPI	2019	2020	2021	2022	2023
CO_2 emissions, tons, scope 1	321	243	265	196	144
CO_2 emissions, tons, scope 2	1.073	942	1.084	866	604
CO_2 emissions, tons, scope 3	97.903	98.027	114.907	103.175	89.379
Total CO_2 emissions, tons, scope 1, 2 & 3	99.296	99.212	116.256	104.237	90.128
Energy consumption, MWh	8.283	8.306	9.008	7.411	7.073
Materials, easy-to-recycle, %	99.6	99.5	99.4	99.6	99.7
Circular Sales, %	86	86	85	88	89
Gender diversity, in company, %	37	36	35	37	37
Gender diversity, leadership, %	23	21	24	33	29
Gender diversity, BoD, %	33	33	33	33	33
Sick leave, %	6.3	6.7	7.3	7.9	7.9
Accidents, no.	11	5	7	9	4
Employee well-being score	6.8	6.8	6.7	7.1	7.2
Employee well-being, response rate, %	88	91	86	86	91
Customer satisfaction, NPS score	28	47	29	21	42
Customer satisfaction, response rate %	34	42	40	29	42
Ratios	**2019**	**2020**	**2021**	**2022**	**2023**
CO_2e in ton/ton material converted	7.3	7.0	7.2	7.7	7.0
MWh/ton material converted	0.608	0.588	0.561	0.551	0.546
Accidents/million working hours	34.6	17.6	24.9	30.9	14.2

The ratios are calculated as follows:
CO_2e in ton/ton material converted = CO_2 emissions, tons (scope 1, 2 and 3)/ton material converted
MWh/ton material converted = Energy consumption in MWh/ton material converted
Accidents/million working hours = Number of accidents/number of realized working hours × 1,000,000

In 2023, Plus Pack earned a bronze medal in EcoVadis rating, which is a global standard for business sustainability ratings based on 21 sustainability criteria across four core themes: environment, labor and human rights, ethics, and sustainable procurement

Which stakeholder(s) are the drivers of the transition to sustainability in the company: Top management and owners	Which stakeholders are the most affected by the transition to sustainability: Suppliers of raw materials on data and finance on reporting principles

What were the largest barriers in introducing a new business model/product in the company and in the marketplace:
Lack of harmonized legislative incentives across geographies

Author's Comments (Gitte Haar):

Plus Pack is in a challenging industry of single-use food packaging, based on materials that lack recyclability with main impacts in scope 3. Especially the use of plastics and the creation of food safe recycled plastic materials are a challenge. In general, the recycling rate of plastic is improving at a low rate both globally and in the EU, and food contact material is moving even slower toward circular economy. Legislation on food safety and traceability of rPET limits the access to recyclable plastic. Aluminum holds records and traditions of a recycled material. When Plus Pack has chosen circular economy and SDG#12 as the core of sustainability and ESG, it is very ambitious and foresighted. Plus Pack has chosen a partner strategy working along the full value chain to influence the availability of recycled raw materials as well as influencing customers and consumer behavior. This case is a good example of how challenging implementation of circularity is within the material loops that have historically been poor in recycling, as plastic and textiles.

Even if Plus Pack decided to recycle plastics themselves by extruding plastic from post-consumer waste there would not be post-consumer PET available in the marketplace to meet the demand of recycled plastic in the market. The critical part of implementing circular material loops is collecting sufficient volume of PET from the value stream of food packaging materials. It is critical that significant volumes are collected in a traceable and safe way, and this is not the case until the large markets have introduced collecting, handling, and disposal systems of a significant part of the materials. Maybe future technologies and legislation will open the input streams from other uses of plastic for food contact material, but now almost all the collected rPET is used for drinking bottles.

Overall, the recycling of plastic is possible; technology is available but a structural collection, sorting, and standardizing of legislation, as well as safe and transparent material flows, are lacking. All this requires political will. The updated Extended Producer Responsibility (EPR) on packaging materials may be an important door to a circular future of plastic and food contact materials, but the industry and the municipalities, often handling the household waste, must work holistically and engaged to solve handling throughout the value chain. Thus, a partner strategy influencing and facilitating stakeholders in the value chain seems to be the only possible in the short run

Plus Pack might be challenged by the increasing demands for reusable food containers (instead of recyclable) for take-away food that is spreading throughout Europe. It is important to prepare for these new trends that might disrupt the market. Plus Pack is a medium-sized provider in the marketplace and will have time to adopt and will potentially also be more agile to adapt to new market conditions compared to larger providers

Aluminum is an underestimated material for food contact material also suitable for both recycling and reuse. There is potential in pursuing new solutions based on this material together with strategic customers, as the material loop here is closer to circular compared to the plastic loops. Even if the large food providers are large and conservative businesses, it is important to interact continuously to find new and more sustainable solutions and harvesting from a first-mover position. This is only done by continuously being on top of knowledge on the industry, the value chain, the materials, and circularity, as seen here.

3.7   AVL

Aage Vestergaard Larsen A/S
Klostermarken 3
DK-9550 Mariager
https://avl.dk

Aage Vestergaard Larsen A/S
Genanvendelse af plast siden 1972

Products, services, and business model: Production within mechanical recycling of plastic and sale of recycled plastic granules for plastic manufactures **Value Proposition in short:** Plastics should always be recycled with the highest quality, thereby the "same" plastic can be mechanically recycled several times before pyrolysis and/or chemical recycling, and most value is contained Provide the same quality of mechanically plastic granules as virgin granules	**Framework of sustainability:** SBTi and SDG#12, SDG#8, and SDG#17 **Two strategic focus areas for 2024:** • Working environment with a focus on minimizing occupational accidents and illness is a strategic focus area. Occupational accidents (and near-accidents) are registered, reasons are explained, and action plans are implemented to avoid repetition. Illness/absence is registered and treated and prevented in the same way. Work hours of three shifts and time of work hours are also recorded, and it is detected whether some shifts are more exposed to accidents than others • Total reduction of our CO_2 footprint in the full value chain and a corresponding focus on reducing the CO_2 footprint per kilo recycled plastic produced. Reducing production waste and electricity consumption are the main focal points in 2024

Short description of the positive impact that your sustainability work has created on the planet and the people by products, deliveries, and the transformation toward a sustainable business:

AVL's sustainable product is 100% recycled plastic granules. The plastic granules are the raw material to produce plastic products
The granules are primarily provided to the Danish and Swedish plastic industry, which uses the recycled plastic as an alternative to virgin plastic. Our focus has always been, and still is, to be able to produce a 100% recycled plastic raw material (granules) for the European plastic industry in a quality that can replace virgin plastic granules. Most recently, we have developed a technology that makes it possible to recycle sorted household plastic waste with high quality that it is approved for cosmetics and hygiene products, which is the next highest hygiene standard required for the packaging products. Hygiene products are examples that it could be used for packaging liquid soap, shampoo, or conditioner. The highest hygiene standards are food contact products

Sustainability certificates and ecolabels that the company uses:
ISO 9001, ISO 14001, Operation Clean Sweep, EUCert

Sustainable Value Proposition/Customer Impacts: By using AVL raw material—recycled plastic—our customers save more than 2.4 kg CO_{2e} per kilo of recycled plastic used compared to virgin plastic (Scope 3 savings)	

Revenue (2022/2023 full year) (€): 2023: >10 mio. € **Number of Employees (2022/2023):** 2023: 60 **Scalable potential:** The demand for recycled plastic granulates is huge and meeting the market demand requires investments in new production lines and the workforce to handle the processes. Theoretically AVL could be scaled 10 times if investments and workforce were in place	**SDG #12. Circular business model:** Plastic waste both pre-consumer and post-consumer including household waste is purchased—It is sorted, washed, and processed into granules, which is the raw material for the plastic industry. Granules are primarily sold in Denmark, and 30% exported to northern European plastic product manufacturers. The AVL-developed quality is an alternative to virgin plastic raw material, whereby the significant CO_2 reductions are achieved when using the recycled plastic as raw material
Ready for the Green and Circular Economy since: Since 1972, as recycling of plastic has been the core business of AVL and AVL is born into the circular economy	**Sustainability Take-off position:** That plastic waste was sent for incineration in 1972, and our mission and vision have always been that plastic primarily must be recycled mechanically
Biggest achievement with sustainability in the marketplace (describe significant business impacts from strategic and integrated sustainability): We have succeeded in processing the sorted plastic waste from households to a quality that this plastic can be recycled and used for products again and again. This closes the loop of plastic waste from households which has been a challenge and confirms the importance of sorting waste at household level	**Significant impact measures (non-financial/ ESG) achievements on environmental, social, and governance impacts:** As the industry's leading partner, we create long-term value for our stakeholders • We make recycling possible more than once • We retain and develop highly competent and experienced employees • We develop sustainable solutions in a circular economy together with our stakeholders to ensure that waste becomes a valuable resource and thus a reduction in global CO_2 emissions Examples of projects/cases where there have been major CO_2 gains for the environment: • The technology to process these difficult plastic products has been developed in close multi-year collaborative projects with the entire value chain. From 2018 to 2020, the environment has been saved by more than 7 million kg CO_2 by recycling discarded bins • Recycled plastic from 500,000 refrigerators results in a saved CO_2 emission of more than 1 million kg
Which stakeholder(s) are the drivers of the transition to sustainability in the company: It has always been AVL's core philosophy that it is a state rather than a transition. In other words, with AVL everyone is involved in the core. Board together with management has driven the development of AVL	**Which stakeholders are the most affected by the transition to sustainability:** Employees Customers Value chain

What were the largest barriers in introducing a new business model/product in the company and in the marketplace:

Slow legislation on ownership of waste in the various collection stages. The necessary legislation has taken 10 years to implement! This has prevented us from building larger pre-sorting facilities for plastic waste in DK, which has led to a significant export of Danish plastic waste

The plastic manufactures procure at the cheapest prices and not the lowest carbon footprints. Thus, the price of virgin raw materials determines the price of recycled plastic, since there is no reward for using the most environmentally friendly raw materials, or taxes on using virgin raw materials. This means that sales of recycled plastic are very dependent on the fluctuating prices of virgin raw materials

Author's comments (Gitte Haar):

The recycling industry is often the facilitator of the circular economy and typically possesses deep knowledge about the materials they are recycling. In the Nordics, waste incineration has been a large part of waste management, which is why the recycling industry has not been able to make profit from recycling of plastic. As input of plastic to incinerators has been important for district heating especially in Denmark, Sweden, and Norway.

This case is important because AVL is in the forefront of recycling plastic mechanically and now also from plastic from household waste with high quality and safety standards. It is important to showcase the plastic recycling companies as plastic is one of the materials that needs investment, technology, and business scaling. Building the infrastructure of mechanically recycling of plastic is very important, because historically it has been lacking. Plastic can become an eco-friendly and low GHG impact material when it is fully recycled and it has almost infinite recycling potential, compared to renewable materials as fibers that will reach an end after few cycles. Plastic is still a larger polluter of nature due to its spilling into nature. Compared to metals and fibers, there is a genuine need for showcasing the companies that make a business of mechanical recycling and can meet and create market demands. AVL is now able to provide numerous qualities and standards, except food contact material-proof plastic. This is due to many years of building experiences, knowledge, and investments in R&D making this case trustworthy. This case is important and hopefully the business will scale over the next few years with the increasing market of recycled plastic of the new qualities, as well as the increasing requirement for recycled and recyclable plastic by industry and legislation.

3.8   KALK

KALK A/S Bredeløkkevej 12 4660 Heddinge Denmark https://kalk.dk/	NYBYG **KALK**® – Build lasting culture
Products, services, and business model: Production, distribution, and sale of lime mortar and lime paint for the construction industry and retail. Lime mortar can be used for both renovation and new constructions When bricking with KALK's hydraulic lime mortar, both stone and the mortar can be reused and recycled. The stone can be reused directly after cleaning, and the mortar can be used in the production of new lime mortar **Value Proposition in short:** Develop and offer lime hydraulic and lime-based products for renovation and new construction based on a circular business model where both mortar and bricks can be recycled and reused. The products are free of harmful chemicals and contribute to a good indoor climate	**Framework of sustainability:** Cradle2Cradle and four SDGs that are an active part of KALKs business strategy (see below)

Selected impact areas:

SDG#6 Clean water and sanitation: We have our own water treatment plant ensuring that we do not discharge contaminated wastewater from sanitation and production

SDG#7 Clean and sustainable energy: We have installed two wind turbines that produce more electricity than we consume. We have biogas in our plant and LED lights in all halls as well as office buildings

SDG#8 Decent jobs and economic growth: We aim for growth, only in the phase where our employees can keep up. If a new machine can make the work easier for our employees, and at the same time producing more and faster, and saving energy, then we will do so

SDG#11 Sustainable cities and local communities: We work for everyone to build solid structures with lime. Buildings that last for a long time, be healthy to live in, and where materials can be recycled. We are part of Stevns World Heritage to preserve and communicate culture locally

SDG#12 Responsible consumption and production: We produce from renewable energy, so we can build houses with a clear conscience that can last for many years

***Short* description of the positive impact that your sustainability work has created on the planet and the people by products, deliveries, and the transformation toward a sustainable business:**

Providing circular products based on lime and lime mortar, as well as minimizing climate impact and increasing the potential to reuse and recycle our own materials as well as bricks. See Sustainable Value Proposition.

Sustainability certificates and ecolabels that the company uses (both product labels and company certificates):

All products Cradle2Cradle certified at gold level.

Selection of five Sustainable Development Goals that we actively worked with.

European Chemicals Agency (ECHA)—the REACH Regulation certification.

EPD—Environmental Product Declaration (necessary for providing materials for certified buildings according to DGNB, BREEAM, LEED).

Sustainable Value Proposition/Customer Impacts: 30% less CO_2 emissions from the choice of lime mortar over traditional cement. Circular material that can be separated and recycled which cement cannot. If we are to recycle the bricks again in 100 years, we will have to use lime mortar. Healthy indoor climate. The limescale absorbs moisture in the room and is a good stabilizer for a healthy moisture balance and indoor climate	
Revenue (2022/2023 full year) (€): Not published **Number of Employees (2022/2023):** 7	**SDG #12. Circular business model:** Our vision is that innovation, knowledge, and open partnerships can pave the way for sustainable development and green transition in the construction industry toward circularity. We believe that the materials that have been the basis of our buildings for centuries again must be at the center of this development. There is too much material waste in the construction industry. With solutions based on sustainable production and recyclable materials, we can meet the green transition. We need to know our past to develop the future in a sustainable direction. There is a lot of knowledge to be gained if you look historically. KALK provides lime mortar that is recyclable and that facilitate the reuse of bricks again and again
Ready for the Green and Circular Economy since: Our materials have always been suitable for reuse and recycling, and since 2014 we actively transformed the company according to Cradle2Cradle and the six SDGs to meet the requirements of the future	**What initiated the strong focus on sustainability, ESG, or SDG:** KALK wanted to expand our business and developed products for new buildings, as we mainly delivered for renovation and restoration of historical buildings. The circular and healthy properties of lime materials need to be implemented in the new construction industry to create circularity and get rid of construction waste
Biggest achievement with sustainability in the marketplace: KALK increased revenue from the green transition and has entered new strategic collaborations. KALK has introduced new painting products to the market based on lime and with no chemical additives called REN (CLEAN) that are suitable as indoor as well as outdoor paint. Market entry of a new product B2B as well as through retail is a big task and requires large investments and still procurement does not put enough value on sustainable and clean products	**Significant impact measures:** GHG savings of 30% in scope 3 Recyclability of our products Net-positive GHG in scope 1+2 Clean and healthy indoor climate in new buildings

Which stakeholder(s) are the drivers of the transition to sustainability in the company:	Which stakeholders are the most affected by the transition to sustainability:
Partnerships in the building industry and customers	Customers and employees

What were the largest barriers in introducing a new business model/product in the company and in the marketplace:

Transparency in production is good. Unfortunately, ensuring transparency is expensive, especially for smaller companies like KALK. Certifications and thorough checks of the value chain are time-consuming. It is also difficult to operate with the reuse of bricks, as it is very cheap to acquire newly produced. If bricks had the same value as gold, we would only reuse and recycle bricks. There is a need to look at whether materials are too cheap, and why the industry recycles and reuses so little.

Unfortunately, there are also challenges at political level. Politicians make fine speeches, but they rarely include sustainability in their own choices. They simply don't set a high enough standard by not leading the transition through public procurement. It is still expensive to create a real take-back system that is economically viable for lime mortar and for bricks. Technology is constantly improving and becoming cheaper, but there is still some way to go before it becomes a good financially sustainable business

Author's comments (Gitte Haar):

KALK is an impressive, old, and small company that by the values of old building traditions and materials is revitalizing the building industry and making materials reusable and recyclable again. They have reestablished their brand and market position and are pushing the industry. This is a very good example that a lot of good solutions are available if we look back, especially in the building industry, but also in other industries. We do not have to invent a lot of technologies; in many cases, we just need to reinstall existing knowledge on materials and methods that has been discarded in the linear economy and the hunt for cheap prices.

KALK has done a good job in re-introducing lime mortar for new buildings and moving out of the old perception that lime mortar is only for renovating very old buildings and churches. They are also pushing the larger manufacturers by their strong branding and that is important. The introduction of the ecolabel and concept of Cradle2Cradle was an important part of framing of existing sustainable products and solutions in new ways. KALK has also taken the long road of getting their products and processes fully documented to prove sustainability which hopefully will pay off. KALK has used the SDGs to communicate their impacts and that has proven strong in the building industry. This is a good showcase of an SME that has used time, money, and efforts in working strategically with the sustainability frameworks available.

3.9   FabricAir

FabricAir A/S
Sandvadsvej 2
4600 Køge
www.fabricair.com

Products, services, and business model:
FabricAir Group designs, develops, engineers, manufactures, and markets fabric-based
ventilation systems for air distribution. Products are each year sold in 120+ countries. The
primary portion of sales are handled through its own legal entities—and the rest through
distributors and partners. Typically, products are sold to Mechanical Contractors who use the
FabricAir Dispersion Systems as part of their total system delivery to the end user. End users
represent a variety of different industries. All sales are business-to-business. In selected
countries, FabricAir Group also offers the installation of our systems as a subcontract to the
Mechanical Contractor.

Value Proposition in short:
FabricAir offers global, innovative, and trustworthy HVAC solutions
Climate change and its effects are the challenge of our time. We develop products and services
that enable energy savings and the avoidance of greenhouse gas emissions when the products
are used. We strive to reduce the negative impact from our own operations and along the value
chain.

Framework of sustainability:
FabricAir has chosen SGDs as the reporting framework and is working on implementing the
EU Sustainability Standard on ESG (CSRD) next year and connecting the legislative
framework with the SDGs. The double materiality assesment identified E1, E2, E4, E5, S1-4
and G within scope of the reporting and strategic implentation i FabricAir.

Selected impact areas:
The chosen SGDs are:
SDG 3: Good Health and Wellbeing
SDG 5: Gender equality
SDG 7: Affordable and clean energy
SDG 8: Decent work and economic growth
SDG 12: Responsible consumption and production

***Short* description of the positive impact that your sustainability work has created on the
planet and the people by products, deliveries, and the transformation toward a
sustainable business:**
FabricAir has started the transition toward circular business models to minimize resource
consumption in the value chain and to minimize climate and environmental impacts in the
value chain. The stepping stone was the acquisition of BorealisWind in Canada providing a
special Ice Protection System, installed inside the blades of wind turbines. BW Solutions are
introduced to the market as a System-as-a-Service, since it not only provides products but also
is a data-driven solution. Learnings from BW and this product line will influence the transition
of other product lines of FabricAir.
The FabricAir Code of Conduct describes the fundamental principles related to ethics and
social and environmental performance. All employees, business partners, and the Board of
Directors are expected to follow these principles.

Sustainability certificates and ecolabels that the company uses:
ISO14001, EPD, and CSRD

Sustainable Value Proposition/Customer Impacts: Facilitating solutions involving employees, customers, and suppliers to meet a responsible and sustainable future, minimizing the impact on the planet, and maximizing the positive social impacts	
Revenue (2023 full year) (€): Gross Profit 2023: No disclosure **Number of Employees (2023):** YE 2023: 214	**SDG #12. Circular business model:** After FabricAir took over BorealisWind in June 2023, the remaining part of 2023 was spent to change the business concept completely. As of January 1, 2024, the solutions are primarily offered to the market through a "System-as-a-Service" concept. The advantage of this concept is that the customer does not have any investments (Capital Expenditures/ CapEx) related to implementing the Ice Protection System— instead the customer pays an annual fee for the use of the system (Operating Expenditures/OpEx). Except the off-balance solution there are other advantages. The new concept gives FabricAir a unique possibility to be in control of the whole life cycle of our product and thereby offering a circular business model with minimum waste throughout the life of the system. The system incorporates the traditional FabricAir technology to distribute the air and is based on advanced sensor technology to identify ice before it builds up on the blades. If an icing event is about to happen, the system switches on and heats the blades thereby preventing icing from affecting turbine performance. The BorealisWind S-a-a-S circular business model is illustrated here: 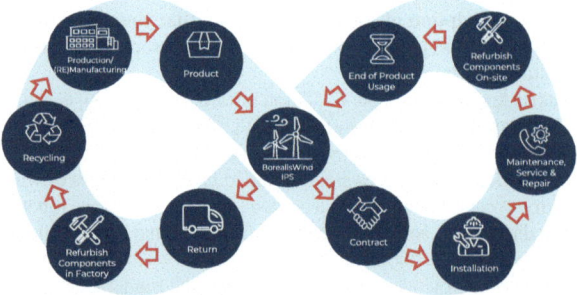 The circular business model is slowly introduced to the products within FabricAir by designing products modularly to be able to repair and maintain the products with customers, and by creating full documentation and traceability of products and materials. Future take-back solutions from customers are evaluated

Ready for the Green and Circular Economy since:	Take off position:
FabricAir sees sustainability as a business imperative and since 2022 management has worked with sustainability. Now sustainability is a pillar in company strategy. Implementation and enrolment of SDGs and ESG is a continuous process and will as of 2024 be business integrated	General understanding of responsible business conduct as the basis for scaling the corporation and participation in training on the green transition for companies in 2022 that kicked off the work on environmental impacts and circularity. In 2023, sustainability became an integrated part of the new strategy
Biggest achievement with sustainability in the marketplace:	**Significant impact measures:**
The interest for BorealisWind Icing Protection solutions as a SaaS is receiving large market attention The ecolabeling of products in the building industry is demanded and the introduction of product LCA and EPD to the market in 2024	Since 2022, FabricAir has disclosed its GHG emission data in the Annual Report including data back to 2019. As of 2024, the EU ESG framework will be introduced in the Annual Report, as well as part of management reports, based on CSRD legislation and the European Sustainability Reporting Standard (ESRS) FabricAir has a top-down approach to sustainability and the executive management team has during 2024 worked actively with building a sustainability roadmap for FabricAir to become climate neutral and circular. Implementing sustainability with top management has resultated in strong targets of climate neutrality in scope 1+2 in 2027 and full climate neutrality in 2030 which requires a full circular business model across all products.
Which stakeholder(s) are the drivers of the transition to sustainability in the company:	**Which stakeholders are the most affected by the transition to sustainability:**
Employees and the new leadership team (EMT) are the drivers of the transition Customers are important in understanding the future and suppliers are essential to drive change in the full value chain	All stakeholders are affected as it is a change of business conduct. The largest barrier is the general lack of knowledge within and outside the organization

What were the largest barriers in introducing a new business model/product in the company and in the marketplace:

Financing a circular business model and SaaS is a challenge.

The lack of transparency in documenting and understanding sustainable and ESG impact in the market still allows for greenwashing on both company and product levels

Author's comments (Gitte Haar):
Being part of the building industry, the market is requesting new sustainable solutions also for FabricAir. FabricAir is a company that sees sustainability as a competitive advantage in a market with increasing requirements for sustainability but also with lots of greenwashing. FabricAir has decided to use sustainability as a strong driver for scaling the company globally as well as developing new business models to close the loops of a material that is still challenging to recycle, namely textiles. Recycling of polyester fiber-to-fiber is still immature, and it is necessary that global actors like FabricAir show the ambition and the way to become fully circular. This is a task that involves the full value chain and potentially also new suppliers and new technologies, as well as the development of take-back systems and collaborations with customers and recyclers locally. Only by seeing sustainability as a business imperative the building industry will minimize its ESG impacts and become circular as required in the EU and other regions. The ambitions and the introduction of the new de-icing system by BorealisWind prove that this is at the top of management's agenda. When successful this will be a cornerstone in scaling the company globally and a showcase to be followed.

3.10   Valtavalo

Valtavalo Oy Head office: Sammaltie 14 90620 Oulu, Finland https://valtavalo.fi/	

Products, services, and business model:
Valtavalo's concept is based on a standardized (EN62776) replaceable light source, a LED tube, which can be easily installed to an existing fluorescent luminaire or to a completely new luminaire designed for LED tubes. A luminaire with a replaceable light source is an ecological and high-quality solution that saves money and nature.
Valtavalo produces LED tubes in their own production plant in Finland. Local manufacturing is enabled by a high degree of automation in production.
In addition to its own sales organization, customers are served through wholesalers and distributors throughout Finland and Europe.
Production and logistics facilities in the city of Kajaani, Finland. Branch office in Sweden.
Value Proposition in short:
Valtavalo Oy is a Finnish company specialized in energy-efficient and environmentally friendly lighting solutions with a focus on replaceability, flexibility, and long lifetime

Framework of sustainability:
n/a
Selected impact areas:
Environmental and financial impacts; our solution saves nature and money

***Short* description of the positive impact that your sustainability work has created on the planet and the people by products, deliveries, and the transformation toward a sustainable business:**
– The solution saves a significant amount of energy and CO_2 emissions compared to older lighting technology
– Long lifetime of the lighting, possibility to **reuse, update, repair, and recycle of products, instead of always replacing the products with completely new ones, and thus avoid unnecessary consumption of materials**
– This also means significant savings in maintenance and upkeep costs in lighting
– The contribution of high-quality flicker-free light to the working environment—well-being and safety of employees
– Customers have an easy possibility to get rid of old toxic mercury-containing fluorescent tubes

Sustainability certificates and ecolabels that the company uses:
Valtavalo's operations are certified according to the ISO 14001:2015 environmental management system standard
Valtavalo has also been awarded EcoVadis Silver (CSR Rating Silver) by EcoVadis, the international organization that assesses corporate responsibility, for our work on sustainable development. Our overall rating places us in the top 25% of the EcoVadis rating. The proud work for sustainability over the years will continue now and in the future, since there is always room for improvement. It is important that the offered solutions last over time, reduce unnecessary material consumption, and thus help save the environment

Sustainable Value Proposition/Customer Impacts:
– Significant documented energy savings compared to traditional lighting technology
– The customer also significantly reduces their carbon footprint
– The customer avoids unnecessary consumption of materials. The replaceability of the light source makes it possible to reuse the lighting unit for many more years
– The customer has an opportunity to reuse, update, repair, and recycle products

Revenue (2022/2023 full year) (€): 3.1 M € **Number of Employees (2022/2023):** 16	**SDG #12. Circular business model**: *User-friendly, maintenance-free, and energy-saving long-life lighting solution with replaceable light source

SDG #12. Circular business model:

*User-friendly, maintenance-free, and energy-saving long-life lighting solution with replaceable light source

*Modular and standardized light source allows the customer to service/upgrade the lighting system by replacing only the light source instead of the whole luminaire or lighting solution (luminaires, wiring, control systems)

*Optimized manufacturing in Finland with low product/material waste using regional energy production and robots. Our products are made of easily recyclable materials (aluminum, PC plastic, and electronics). We use as much as possible local suppliers. We recycle packaging materials from manufacturing materials into packaging materials for finished products sent to customers

*The EU Ecodesign Directive from 2021 encourages companies to engage in the circular economy and draws attention not only to energy efficiency but also to the opportunities to **reuse, update, repair, and recycle products**. The new regulation also encourages replaceability of light sources and separate control gears. They all contribute to the beneficial use of materials and thereby reduce emissions and adverse environmental impacts of the product or service. All these aspects are already taken into account in solutions with replaceable light source.

• **Replaceability of light source:** in our solution, the light source can be easily replaced without tools and without the need for special skills

• **Long lifetime:** LED tubes made in Finland have a long lifetime

• **Upgradability:** lighting can be updated or upgraded easily by replacing the light source in the luminaire or by increasing/decreasing the number of light sources

• **Repairability:** the maintenance is as simple as replacing the light source. LED tube contains all the technology needed to create high-quality lighting. The luminaires, on the other hand, have no separate breakable chokes or ballasts and are maintenance-free

• **Reusability:** light source can be flexibly moved from one luminaire or space to another as required. Also, LED tube is a standardized light source and continuous availability is ensured by compliance with the international IEC 62776 standard, published in 2014

• **Recyclability:** at the end of their long life cycle, our products can be recycled

Ready for the Green and Circular Economy since:	Take-off position (What initiated the strong focus on sustainability, ESG, or SDG):
Valtavalo's operation has been based on the circular economy since the company was founded (2008). We are pioneers and advocates of sustainable lighting. It is important for us to be able to provide lighting that withstands the test of time and is environmentally friendly	n/a
Biggest achievement with sustainability in the marketplace:	**Significant impact measures (non-financial/ESG) achievements on environmental, social, and governance impacts:**
Awareness of the environmental effects of lighting and the demand for the lighting solutions we offer are increasing. More and more companies find the environment and responsibility increasingly important	n/a
Which stakeholder(s) are the drivers of the transition to sustainability in the company (management, project managers, etc):	**Which stakeholders are the most affected by the transition to sustainability (management, customers, employees, suppliers, etc.)**
The owners and the management of the company	n/a

What were the largest barriers in introducing a new business model/product in the company and in the marketplace:

For a long time, Valtavalo were the only ones conveying our message (replaceability of light source), and it is therefore particularly wonderful that the European Union has finally realized the importance of the issue. The EU's Ecodesign Directive that was implemented in the autumn of 2021 encourages companies to engage in the circular economy and draws attention not only to energy efficiency but also to the opportunities to reuse, update, repair, and recycle products. All these aspects are already considered in our products.

During the first years there was hardly any competition as LED tube was a new invention. In December 2014, a standard for LED tubes (EN62776) was completed that guarantees a safe replacement of the traditional T8 fluorescent tube. Thanks to the standard, it is possible to replace old fluorescent tubes with LED tubes and use standardized LED tubes regardless of brand.

Author's comment (Gitte Haar):

This is a showcase of circular economy in an industry where the linear economy has dominated since the introduction of LED. The norm became that LED light source was fully integrated into the light-fitting and thereby both the fitting and the LED light source must be changed when the LED is worn out. This crazy approach must be changed to save resources and critical raw materials. Valtavalo is one of a few providers of solutions delivering the light source into existing fittings. A solution that should be the norm rather than the circular exception. Valtavalo is showcasing the solutions of the future and pushing large providers such as Phillips and Ikea and therefore this is an important case.

3.11   OnePan

OnePan AB Bögatan 1A 412 72 Göteborg https://onepan.se	**ONE**Pan®
Products, services, and business model: OnePan markets, sells, and manufactures re-coatable non-stick kitchenware, with ceramic PFAS-free coating **Value Proposition in short:** In a world where toxic non-stick kitchenware is bought, used, and discarded all within few years, OnePan challenges the norm OnePan is built on the radical idea that one pan is all you'll ever need. Does that mean that you should accept food sticking for the rest of your life? Not at all. As chefs, we understand that busy kitchens demand uncompromising performance. We simply do what others say they do when they don't—circularity, for real	**Framework of sustainability:** SDGs **Selected impact areas:** SDG #3—Good health and wellbeing (No to PFAS—read more about PFAS and our coating here) SDG #6—Clean water and sanitation (No to PFAS) SDG #9—Industry, Innovation, and Infrastructure (Provoking change in the industry, leading the way for circular innovation) SDG #12—Responsible consumption (from linear to circular) SDG #13—Climate action (The production and distribution of a new OnePan emits approx. 30 kg CO_2. A re-coat (full process from consumer back to consumer) emits 1.5 kg, i.e., 20 times less than new production. Also important is that a new OnePan is arguably more sustainably produced within the EU than other China-produced budget non-stick pans) SDG #17—Partnership for the goals. In order to create structural change, we need to work together and join hands with other like-minded organizations. OnePan works with RISE, Chalmers, HoloHouse, and Cradlenet among others to drive change

Short description of the positive impact that your sustainability work has created on the planet and the people by products, deliveries, and the transformation toward a sustainable business:

Please refer to the LCA calculations in SDG #13 above.

Instead of having to throw away a perfectly intact product, just because the top layer is worn out, we provide a solution that will save both money and CO_2 emissions for the consumer. We also provide an example for the industry and inspiration for other companies to follow. We also lobby hard against the use of toxic PFAS substances—and naturally don't use any of it in our production which minimizes pollution

Our products are built with a strong and robust design so that the top layer can be re-coated, instead of discarding the whole pan, repeatedly. The robust design also gives the products incredible frying properties. Because you won't change a system and sell sustainable products if they are only sustainable, they also have to be qualitative, and people have to want to use them

Sustainability certificates and ecolabels that the company uses:
BRC Consumer Products

Sustainable Value Proposition/ Customer Impacts: Please see above	
Revenue (2022/2023 full year) (€): Approx. 600,000 € expected 2024 is 1.5 Mio € **Number of Employees (2022/2023):** 6	**SDG #12. Circular business model:** Here is how our circular model works for the consumer: https://onepan.se/en/pages/circularity Once we receive the worn-out pan to our warehouse in Mariestad (soon Gothenburg), Sweden, we take off the handle, which is made of durable stainless steel, and wash and polish it ready for the next user. Then we send the pan body to Småland in southern Sweden where we blast off the remainder of the old coating. Then we re-coat it and polish the bottom so it looks and feels almost as new, and definitely works just as new. Then it is sent back to the warehouse where a reused handle is screwed back on and packaged ready for the next user ordering a re-coat on the website
Ready for the Green and Circular Economy since: We started in 2020	**Sustainability take-off position:** The founders, two established chefs, didn't see the logic in buying and throwing away expensive frying pans that were almost fully intact, only because the top layer was worn out. It felt wasteful and an unnecessary burden on the environment and the user's economy. They also felt passionately about removing the use of PFAS, with its many alarming side effects on nature and the health of animals and humans, in kitchenware. They didn't want to season their guests' food with poison or contaminate the water we drink and air we breathe with PFAS Therefore, OnePan innovated a substitute to toxic non-stick kitchenware that gets bought, used, and discarded all within a few years
Biggest achievement with sustainability in the marketplace: Getting Sysco (Menigo), a massive global wholesaler for industrial kitchens, to create a white label series of OnePan and offer it to all of their customers in Sweden and hopefully soon worldwide	**Significant impact measures:** See above
Which stakeholder(s) are the drivers of the transition to sustainability in the company: The full team is very engaged, naturally with an inspiring management team leading the way Also, our customers and consumers Influencers help us drive change too	**Which stakeholders are the most affected by the transition to sustainability:** Consumers and suppliers. It's challenging for our production to change from linear to circular mindset, but we are working closely with them to implement circularity

What were the largest barriers in introducing a new business model/product in the company and in the marketplace:
High production prices (it's costly to produce a product that should be durable enough for a circular system), which means our products must be expensive and we have compromised margins to make further investments. Also changing consumption behavior is a secondary challenge, but it gets easier and easier.

Author's comments (Gitte Haar):
OnePan is a very good example of a successful startup evolved out of passion for abolishing waste and creating a sensible solution to avoid overconsumption. We have cases like this that bring specific industry and consumption knowledge to the circular transition and the business environment. OnePan is a good example of being able to scale a very good idea to the market through a good market entry strategy. This showcases the need for transition in every corner of our society, consumption, and product range.

3.12   KLS Pureprint A/S

KLS PurePrint A/S
Jernholmen 42A
2650 Hvidovre
https://klspureprint.dk/

Products, services, and business model:
KLS is a manufacturer in the graphic industry producing packaging materials, print matters, and magazines. Products are sold directly to end-users or through distributors.

Value Proposition in short:
The offset printed matter and packaging materials are produced climate-neutral, Cradle2Cradle, and FSC certified. The packaging materials are certified compostable (EN 13432). This makes it outstanding within the printed packaging materials.

Framework of sustainability (Sustainable Development Goals—SDGs, ESG, SBTi, B-corp, etc.:
KLS has chosen SDGs and the Cradle2Cradle concept for communicating sustainability and transition. As of next year, KLS is subject to CSRD as part of the Bording Group and will then use the European Sustainability Reporting Standard (ESRS)

Selected impact areas that are chosen as strategic important for the company based on the thinking of double materiality—environmental, social, and government impacts (ESG), SDGs, as well as financial impact:
The chosen SGDs are:
SDG#3—Health and Wellbeing
SDG#12—Responsible Production and Consumption
SDG#15—Live on Land
SDG#7—Clean and Affordable Energy

Short description of the positive impact that your sustainability work has created on the planet and the people by products, deliveries, and the transformation towards a sustainable business:

KLS is part of an international collaboration **print-the-change** important for the development of new materials and sustainability in general: https://printthechange.coop/ and by joining and developing this collaboration with other printers, also Cradle2Cradle certified, in a highly competitive industry the joint innovation, product development, knowledge sharing, and cost sharing have been essential.

All actions start with mapping a baseline and then developing a strategy to minimize or completely phase out harmful or polluting elements. This process has often involved the entire value chain from suppliers to customers, and building a systematic framework and method is important.

Sustainability certificates and ecolabels that the company uses:
Product brands:
Cradle2Cradle
The Swan (Nordic Eco Label)
FSC
Climate-Neutral based on an industry developed calculator
EN 13432 (EU composting standard)
The company:
Cradle2Cradle
Member of the Global Compact
ISO Certified under 9001 (Quality) and 14001 (Environment)
As of 2025 reporting according to CSRD

Sustainable Value Proposition/customer impacts:
Vision: "The world's greenest printing house".
Climate-neutral printed matter and packaging materials produced without harmful chemicals in accordance with the Cradle2Cradle Standard and with the target of full traceability throughout the paper production cf. FSC.

| Revenue (2023 full year) (€):
8,713,000
Number of Employees (2023):
41 | **SDG #12. In case of a circular business model describe business model, take-back system, and the new material value chain:**
We have ensured that no matter which material loops our printed matter and packaging materials go through, there are no parts of our product that must end up as waste. This applies regardless of whether the product is reused, the fibres are recycled, or the product is composted.
We have been able to change food packaging (food contact material) from being a material stream that is incinerated to one that can be composted and contributes to a cleaner material stream of natural fibres. The company cannot solve this independently, because it is dependent on waste sorting and infrastructure. The first step for a better utilization of resources in bio-based is cleaner material with no critical chemical. Then both the flow of recycling becomes more efficient and with no waste for landfill, and composting will be possible, as it is not today for many printed materials. |

Ready for the Green and Circular Economy since:	Take off position:
The transition started in 2008 and has been a continuous development since with new targets, and this will never stop.	The green conversion of KLS and a circular business model was a survival strategy for KLS in a highly price-competitive industry in a declining market for advertising print materials 10–15 years ago. Now it is the core of KLS. Since KLS has implemented sustainability to develop the new business area for ecofriendly packaging materials.
Biggest achievement with sustainability in the marketplace:	**Significant impacts:**
KLS introduced Cradle2Cradle certified products in 2015 and here KLS also entered the international strategic work with other European printers in the Print-the-Change community. Today KLS is working with the traceability of the wooden raw materials and regeneration of biodiversity in forests. KLS has not decided whether we will use SBTnature or an EU ecolabel to document traceability, responsible forestry, and biodiversity. KLS is also active in work on transforming the local industry area into a full supply of renewable energy with the community of companies at Avedøre Holme. Sustainability is a new ending process that must be integrated in all employees and processes of the company, and new solutions and ESG initiatives must be developed and implemented.	Climate-neutral in 2009 and ban on fuel company cars in 2011 LCA calculations on CO_2 and thus the first climate-neutral printing company in Scandinavia. Cradle2Cradle certificate in 2015 and Cradle2Cradle Gold certificate as of 2022. Bio-compostable packaging materials in 2020. KLS has gained a lot of knowledge from working with the green and circular transition for many years and this has resulted in methodology, as: All actions start with mapping a baseline and then developing a strategy to minimize or completely phase out the harmful or polluting elements. This process has often involved the entire value chain from suppliers to customers.
Which stakeholder(s) are the drivers of the transition to sustainability in the company:	**Which stakeholders are the most affected by the transition to sustainability:**
The board, management, and owners	All stakeholders are affected by the strategy. Customers and suppliers are central in the development of new materials and products. For employees and management, the sustainable strategy means several limitations and adaptations in how we manufacture, and which materials we can use. On the other hand, it also means a clear strategic direction for the company, that makes it easier for everyone to act in accordance with the strategy

What were the largest barriers in introducing a new business model/product in the company and in the marketplace:
One of the biggest challenges as an SME has been to finance the green transition. In practice, this has often ended up being financed either through operations or through participation in green grants programs.
For the market to understand the differences in the ecolabels and the importance of printed materials without chemical hazards. As a frontrunner forming the marketplace and educating customers in sustainability is necessary.

Authors comments (Gitte Haar):
KLS is a company that many years ago started the green transition out of necessity in an industry that was disrupted by digitalization and has been on a long journey towards more ecofriendly products and company and thereby sustainability is part of the business core working actively in transforming an essential material flow of natural fibers that currently are not used optimally in the biological cycle. KLS is an example of how the green transition may strengthen the competitiveness of many SMEs, if they dare think innovatively, and dare to invest massively in the transition. The KLS case shows that a genuine transition is huge and affects the company, the products, the suppliers, and the customers. KLS was transformed from a traditional advertising printing company to a packaging manufacturer with certified products free of harmful chemicals, and circular bioproducts. The journey is long and never ending as sustainability is a moving target with more and more planetary boundaries being exceeded. Many SMEs need assistance with their transition to a green and circular economy; both strategically, technically, and financially to succeed. The business support system and the financial sector must improve and become better at providing this assistance to the SMEs.

Reference

Haar, G. (2024). *The great transition to a green and circular economy. Climate nexus and sustainability*. SpringerNature.

Outliners

4

During the work with this Case Collection, I came by some cases outside the Nordics that are important to showcase because they outline the need for implementation of sustainability in a way and to an extent that others must learn from and get inspired by. The cases are the City of Amsterdam and their implementation and cooperation with society and business on implementing circular economy and Stanford University that showcases the energy transition on city level including science and living labs with scientists and students at Stanford University in impressive ways.

The cases are written and provided by the people responsible for the implementation from the City of Amsterdam and from Stanford Sustainability Utility & Infrastructure. These cases are important because they show large-scale, long-term transitions. They describe the necessity of science and facts, and how detailed and well planned a transition must be orchestrated to become successful. These outliners are also large scale and there important as the great transition to a green and circular economy is a transformation of the whole market and the full society.

These outliner descriptions are inspirational roadmaps not only for other cities but also for corporations, entrepreneurs, politicians, or citizens who want to lead, engage in, understand, or facilitate the transition to a green and circular economy.

The case on Implementation Agenda for a Circular Amsterdam illustrates the importance of:

- A clear and ambitious vision framed as a burning platform.
- Detailed roadmap to bring the vision into implementation.
- Engaging with all actors in society, such as business, entrepreneurs, municipality employees, citizens, NGO, educational institutions, and other governmental institutions such as the Port of Amsterdam.
- Data-driven implementation brings overview, insightful communication, and the ability to set targets, follow up, and celebrate achieved goals.

G. Haar, *Nordic Case Collection on Sustainability and Transition to a Circular Economy*, Springer Business Cases, https://doi.org/10.1007/978-3-031-78638-9_4

- Accepting and communication of the necessity to fail and learn, and update the agenda.
- Alignment with national and international (EU) regulations.
- Acceptance of the continuous and never-ending journey the City is embarking on and already framing the shape of the next plan to follow 2026.

All these elements, a thorough and strong will, and planning will make Amsterdam successful in its transition. A transition to a green and circular economy is a transition on society level and it is driven top-down with full engagement button-up in society. This is the most important task for local politicians, state, and municipalities to roll out and involve all stakeholders. As described in the case our value chains and the impacts on climate, nature, ecosystems, and biodiversity mainly come from what we consume, and the value chains of our consumption at all levels in society are very long and complex. Therefore, it is a transition at all levels engaging all stakeholders in society driven by economy including the pricing of the environmental externalities. The externalities, as environmental and climate impacts, are not priced today, as clearly illustrated in this case. The agenda on the implementation of circular economy on city level must go hand in hand with the implementation of the energy transition and the transition to sustainable transport as described earlier in this book.

The case on Stanford Sustainability Utilities & Infrastructure is an important showcase with proven, implemented, and science-based methodology to inspire and to follow. Stanford University is a full village situated south of San Fransico in California, USA. The transition is not just a transition of a university—it is a proven transition of housing facilities of students and staff, working facilities for the full university staff and students, and infrastructure and natural areas in and around Stanford. The case on Standford Sustainability Utilities and Infrastructure illustrates the importance of:

- Financials and predictions on future risks, costs, prices, and asset valuation have been an important part of the overall decision scheme, which is often overseen by many today. The cost/price levels of today are often seen as normal not realizing the financial effects of doing nothing and the transition of renewable and adapted systems are part of securing the future assets and credit valuation as well as ensuring stable and predictable operation.
- Monitoring throughout the full implementation is emphasized as an important element in the new Stanford 2025 Climate Action Plan. As well as widening the scope to other resources than energy, to scope 3 and the consumption of goods and transport.
- Showcasing the importance of living labs and the close collaboration between academia, operations, and implementation to drive the transition to a sustainable future. The history of transferring knowledge from science to real life seems stronger in the USA than in Europe, and is important to learn from.

Systematic sharing of knowledge and methods, and taking a forefront position on the transition is very important to spread the science on the solutions and implementation of a fair and sustainable living. All universities conduct research and teach science on the state of planet. At Stanford it is their ambition and responsibility to showcase implementation and to bridge the knowledge gap between science on the one side, and society, businesses, and citizens on the other side. Therefore, this outliner case is so important to share. This is a roadmap to be followed by other universities and stakeholders within and around the universities.

4.1   Stanford University

4.1.1   Stanford Sustainability Utilities and Infrastructure

By: Executive Director, **Lincoln Bleveans**
Stanford University, 94305 Stanford, CA, USA

Institution
Stanford University[1] is a world-renowned private research university comprising 7 schools,[2] 15 interdisciplinary institutes, approximately 14,000 undergraduate and graduate students, almost 100 student residences, related student services like dining and recreation, a sponsored research budget of almost US$2.0 billion, and over 10,000 administrative and operations staff.

Stanford's vision includes four themes[3] inspired by the ideas of our community:

- Sustaining life on earth
- Accelerating solutions for humanity
- Catalyzing discovery in every field
- Preparing citizens and leaders for lives of active citizenship

Woven throughout these themes is a commitment to ensuring equity and inclusion in our research and on our campus, embedding ethics across research and education, and engaging with partners beyond our walls to learn from and give back to our local and global community.

From a utilities and infrastructure perspective,[4] Stanford is a small, self-contained, full-service, and vertically integrated city. As such, Stanford operates its own mission-essential electricity and thermal energy system, water and sewer, waste management, civil infrastructure such as roads and bridges, and building automation and energy management systems.

---

[1] stanford.edu

[2] The Graduate School of Business, the School of Law, the School of Medicine, the School of Education, the School of Engineering, the School of Humanities & Sciences, and the School of Sustainability.

[3] ourvision.stanford.edu

[4] sui.stanford.edu

Sustainability Goals and Targets

Stanford's climate action has historically focused on energy and, within the energy space, on scope 1 and 2 greenhouse gas (**GHG**) emissions. As I will show in this case study, Stanford has taken bold steps to **reduce scope 1 and 2 emissions by a remarkable 78% from their peak in 2011** and is working aggressively on the remainder. **Stanford's ultimate GHG goal is now net-zero scopes 1, 2, and 3 by 2050.**[5]

In the meantime, Stanford has "walked the walk" on other sustainability topics for many years, integrating innovations in decarbonization, water conservation, and waste reduction, reuse, recycling, and composting into the very fiber of its campus.

Stanford's climate action planning currently underway[6] remains focused on energy and GHG but, for the first time (and going beyond the confines of its 2050 goal),

1. Includes water, waste, food systems, transportation, and the investments made by its endowment, and scope 3 topics such as supply chain, travel, the built environment, and employee commutation
2. Works across Stanford's decentralized institutional structure, bringing together researchers, students, alumni, operations, our endowment, outside stakeholders, and partners in higher education, industry, and the Silicon Valley and global startup communities
3. Expressly includes climate adaptation, resilience, and environmental and climate justice considerations

More broadly, Stanford seeks to be a **"living lab"** for sustainability, with academia and operations working together to use our campus systems as laboratories for research and technology application. In doing so, we must carefully balance the mission-essential nature of our systems with opportunities to derisk new technologies, business practices, and change management processes.

To paraphrase Meta CEO Mark Zuckerberg, we must "move fast but **not** break anything."

In this connection, it is important to note that Stanford has not yet publicly adopted a specific framework for its sustainability efforts. Instead Stanford focuses on this academic–operations nexus—the living lab—to both discover world-changing sustainability advances in the laboratory and apply applicable advances in its everyday operations. Stanford has received "Platinum" ratings from both the *American Association for Sustainability in Higher Education*[7] and *The Climate Registry*.[8]

---

[5] Our Climate Action Planning process, underway at the time of this writing, may suggest a net-zero GHG goal sooner than 2050.

[6] https://sustainable.stanford.edu/operations/energy-climate/planning

[7] aashe.org

[8] theclimateregistry.org

Sustainability Utilities & Infrastructure group (SUI)
Within the large and decentralized research university/full-service, vertically integrated city that is Stanford, the *Sustainability Utilities & Infrastructure* group (SUI) is the beating heart, responsible for the mission-essential utilities and infrastructure. At the same time, SUI is charged with leading sustainability, adaptation, resilience, and climate justice (as a whole, "sustainability") at Stanford as a whole and making sustainability real in everything we do. That's an enormous challenge but also an enormous opportunity: when we get it right, we can optimize outcomes for Stanford and provide exemplars for the rest of the world.

This case study is focused on a key part of that effort: the decarbonization of Stanford's energy system.

Strategic Approach

It is no surprise that Stanford requires a significant amount of energy for its 1000+ campus buildings: to support its academic mission and the research functions, to support student life and athletics, and to provide cooling to its hospital complex, including a Level One trauma center.

This energy comprises both electricity to power electric loads and thermal energy for heating and cooling. Efficiently managing energy supply and demand, as well as the corresponding GHG emissions, is critical to the university's future.

For decades, Stanford was in the main powered by an onsite, natural gas-fired cogeneration power plant called *Cardinal Cogen*. Cogeneration (also called combined heat and power or "CHP") is a widely used power plant configuration that employs a combustion turbine and electrical generator to produce electricity while also extracting heat from the combustion process. This heat is then exposed to water, turning the water into steam that can be used in both heating and cooling applications. Interestingly, until very recently gas-fired cogeneration has been the default economic and environmental choice for combined heat and power applications.

In the early 2000s, *Cardinal Cogen* was nearing the end of its useful life. Stanford's need for electricity and thermal energy, by contrast, was getting larger due to campus growth, increasing electrification, and hotter temperatures due to climate change. At the time, building a new and more efficient (and likely larger) gas-fired cogeneration plant was the obvious choice: it was proven technology, the climate impacts of natural gas had not yet been widely acknowledged, and the campus's thermal energy distribution system and building systems were already configured for steam. It would have been very easy for Stanford to lean back onto legacy assumptions such as natural gas, cogeneration, and steam.

That's not What Happened, of Course

Stanford began planning for a replacement for Cardinal Cogen around that same time. As a matter of policy, the university recognized the impacts of fossil fuels—including the natural gas used in cogeneration—on climate change. As a matter of economics, the university identified changes in the California energy markets—especially for natural gas—that would likely result in higher and more volatile energy costs. It was time for a new way to power Stanford University.

And, as befits a science- and engineering-driven institution like Stanford, the planning started at the very beginning: the First Law of Thermodynamics!

Energy can neither be created nor destroyed, but only changed from one form to another.

And we drove forward from there, taking nothing for granted. We challenged every assumption, engaged our academic and operational colleagues around every possible challenge and solution, and collaborated with crucial vendors and outside engineering experts. This stakeholder engagement was not just important but vital: the reliable and cost-effective operation of Stanford's energy system is a prerequisite to nearly every activity on campus, from teaching and campus life to delicate and long-duration research to public safety. It had to work.

With this bold approach, it is no surprise that the *Stanford University 2009 Energy & Climate Plan,*[9] leaned resolutely forward:

Climate change is one of the most significant global socioeconomic challenges for our generation, yet it also provides an opportunity for Stanford University to develop innovative solutions and provide leadership through research, teaching, outreach, and the operation of its own campus.

---

[9] https://sustainable.stanford.edu/sites/g/files/sbiybj26701/files/media/file/stanfordenergyandcli-mateplan_11-10.pdf

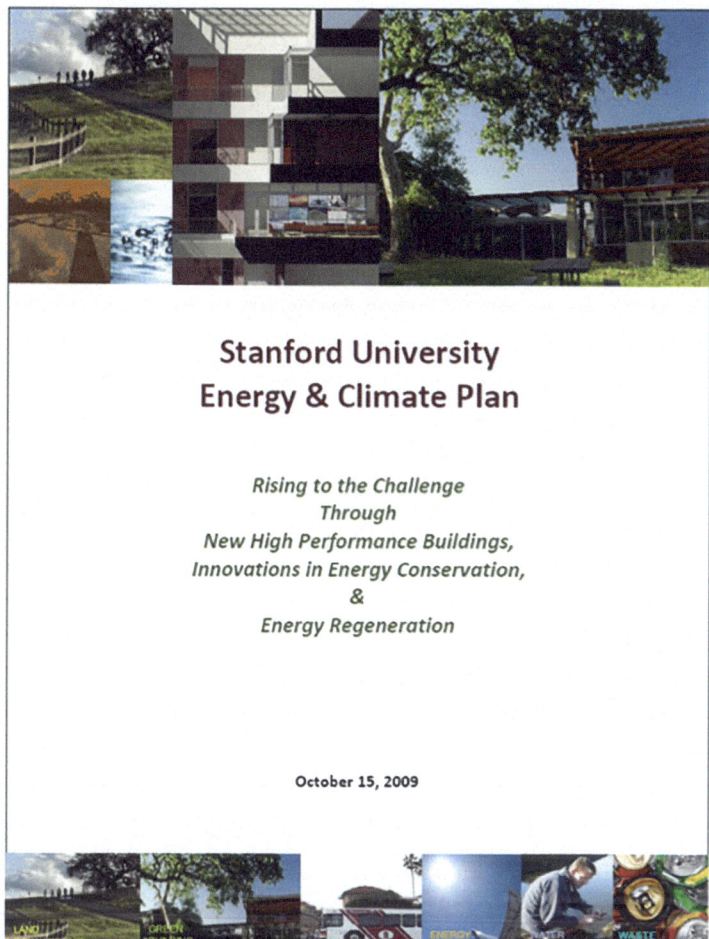

With that in mind, the 2009 Plan laid the groundwork for a fully integrated system from remote utility-scale solar power to heat recovery to switching from steam to hot and chilled water. None of the elements were rocket science: in fact, the technology risk of the system's components was low. The innovation was the combination of those components into the system itself. And at a scale never before attempted in North America. Simply put, the system was the innovation.

cogeneration	conventional + heat recovery chillers
electricity + steam	electricity + heated and chilled water
natural gas	utility-scale solar
California grid backup	California grid dependent

That system, called Stanford Energy System Innovations or "SESI," is a vertically integrated, renewable electricity-powered, heat recovering, thermal energy-storing, campus-transforming, first-of-its-kind at scale district energy system.

Here's what it looks like, component by component:

remote utility-scale renewable electricity **generation**
remote utility-scale battery **storage**
wholesale power **market** participation
local rooftop renewable electricity generation
local hot and chilled water production
local hot and chilled water storage
local hot and chilled water **distribution**
local electricity distribution

And below is how it comes together: utility-scale solar power generation and some battery storage out on the California grid, plus a small amount of local solar, are transmitted to Stanford as a wholesale customer of the California grid. That electricity, representing over 100% of Stanford's electricity needs, provides electricity to campus via the university's own electric distribution system and powers a thermal generation and storage plant called the Central Energy Facility (CEF). The CEF, in turn, sends both hot and chilled water through approximately 50 km of closed loop piping and serves hundreds of buildings: laboratories, classrooms, offices, athletic facilities, Stanford's hospital system, and residential and dining facilities. Those two closed loops are vitally important for both water conservation and capturing the university's 70% thermal overlap.

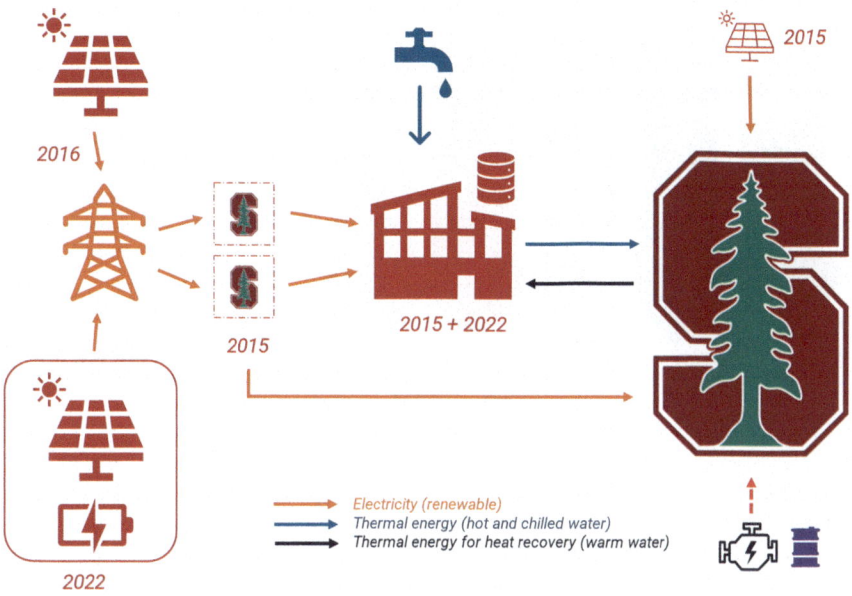

SESI encompasses the best of both North American and European district heating and cooling system advancements, with engineers, manufacturers, and constructors from both continents collaborating on this state-of-the-art transformation of Stanford into one of the most efficient district energy systems in the world. Aside from extremely efficient operations, sustainability features are incorporated throughout SESI so that it can serve as an example for generations to come. The results have been remarkable:

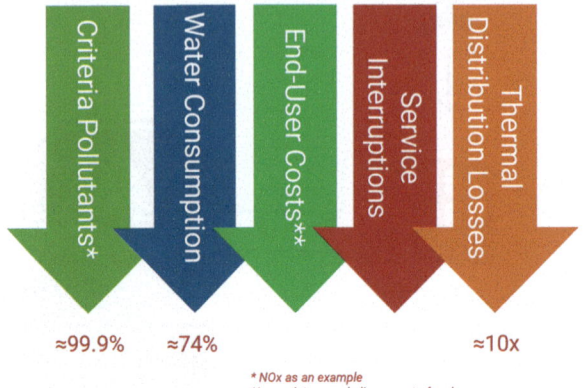

≈99.9%   ≈74%   ≈10x

NOx as an example
** proprietary, excluding a cost of carbon

SESI remains a vivid example of both optimizing for Stanford and providing a template for the rest of the world. At its heart, SESI is a story of opportunity, renewal, and transformation.

Opportunity: Stanford, like many other places, has assets and systems that have reached their end of life. That's an obligation but also an opportunity. The obligation is renewal. The function of that end-of-life asset or system must be renewed. Operations require it. The university must go on.

Renewal: Renewing that asset or system's functionality to satisfy Stanford's operational imperatives, powering the campus in this case.

Transformation: Finding a way to satisfy the *operational* imperative, like powering the campus, while at the same time satisfying the university's *policy* imperatives, like net-zero GHG.

When *Cardinal Cogen* reached its end of life, Stanford had a choice: lean back on legacy operational solutions like natural gas and cogeneration or lean into both our operational and policy goals with something new. In other words, renewal *or* renewal and transformation?

The low-risk option, of course, would have been more of the same. But we leaned in, challenging assumptions and analyzing every possible option. We analyzed the range of possible technology options, from a new cogeneration plant to varieties of gas turbines to heat recovery methods and even whether to remain an autonomous electric system or rely on the local utility, Pacific Gas & Electric, as a retail customer.

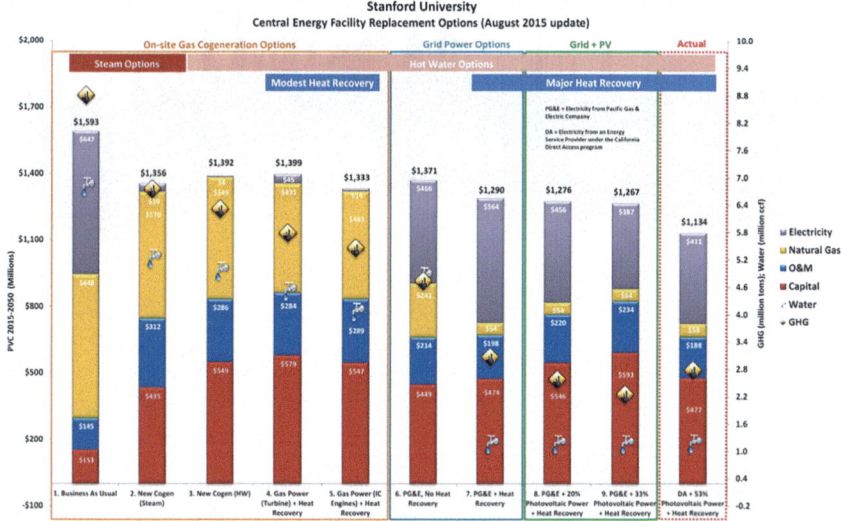

These options were compared for both economic efficiency (on a net present value basis) and climate efficiency (total lifetime GHG). The win-win option, shown in the column farthest to the right, was something quite new. It was the SESI system.

We also analyzed Stanford's thermal energy needs, realizing that heating and cooling needs had a significant overlap to the tune of 70%. Now, instead of waste heat evacuated through a cooling tower, SESI's heat recovery chillers use it as a fuel.

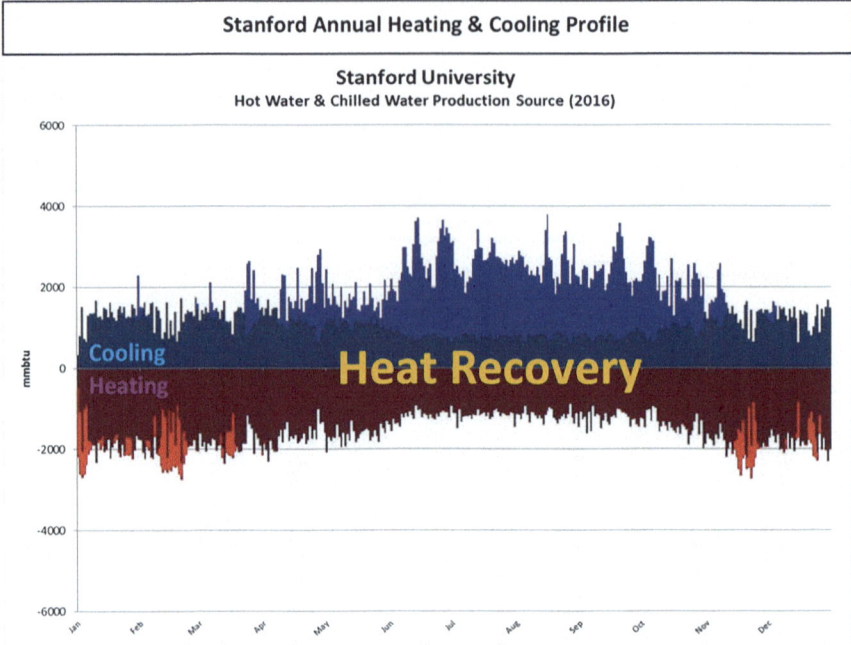

This analysis also showed that our heating and cooling needs themselves could be satisfied with low temperature thermal energy, hot and chilled water, rather than high temperature steam.

Challenging basic assumptions and finding the transformation. That's SESI.

retail → wholesale

fossil → renewable

waste heat → heat recovery

steam → hot & chilled water

≈Ø energy storage → electricity storage and
                                    thermal storage

Roadmap

SESI represents electricity, chilled water for cooling, and hot water for heating, serving nearly the entire full-service city, Stanford. The decarbonization impact has been massive. Look at that solid black business-as-usual line versus the actuals, a *78%* drop in GHG from our peak in 2011.

Stanford Scopes 1 & 2 GHG Emissions Compared to External Benchmarks (1990-2050)

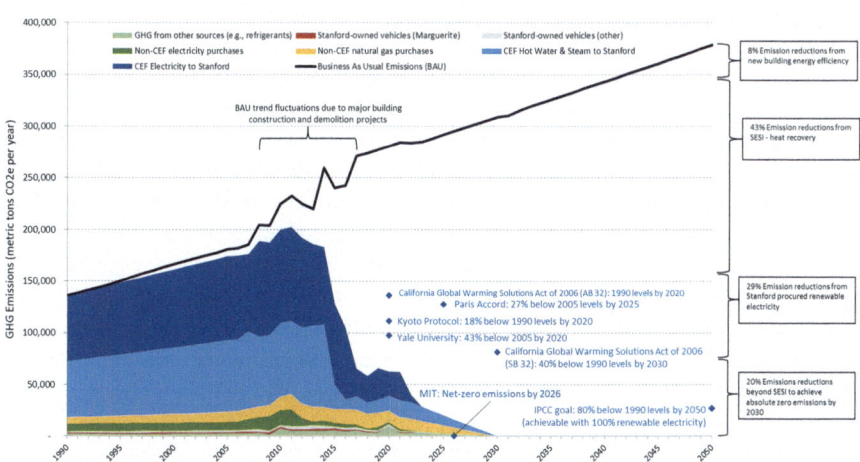

And it's not just carbon, but energy-related water uses down by three-quarters, local pollution emissions down to almost zero, end user costs and reliability far better, thermal distribution losses from the 30% range to the 3% range. Again, massive.

That said, all energy is compromise.

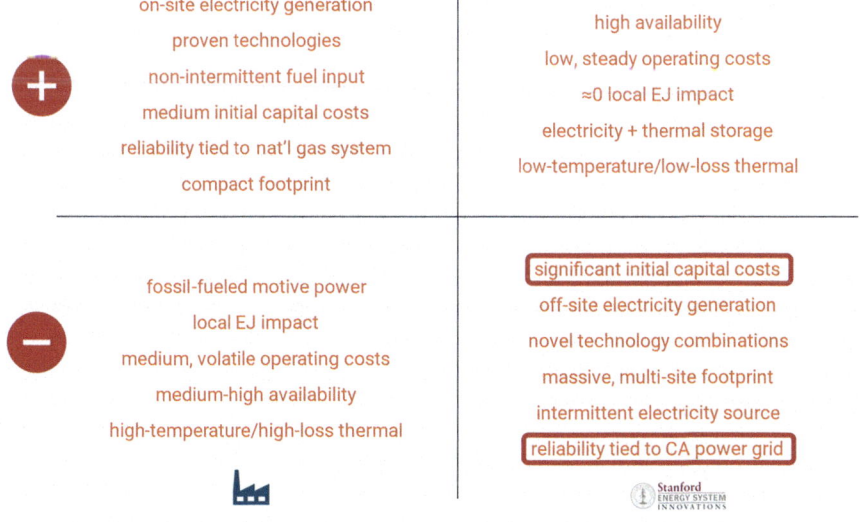

Here are three big ones. First, SESI's design assumptions. When SESI was conceived, the university tasked the engineers with meeting intensively analyzed design assumptions for campus growth and the rate of climate change. Those assumptions seemed prudently conservative at the time, but both were quickly outpaced by reality: the campus grew much faster and got much hotter, far more quickly than expected. The result was chilled water curtailments—the capacity of the brand-new system suddenly insufficient for the hottest summer periods—with two short years (and not decades, as originally expected)—of the system's operation. In response, we made a significant capital investment to nearly double the CEF's chilled water capacity in 2022.

Second, the reliability of this mission-essential system is dependent on that of the California grid. That risk became real during a heat storm in June 2022: a fire underneath our primary transmission line. With only a fraction of our needs available on a back-up line, most of the campus came to a sudden, disruptive, and expensive halt for 3.5 days. Work is currently underway to both harden and engineer redundancy into this grid connection, most likely at significant capital cost to Stanford.

Third, the initial capital cost of a facility like this is quite steep, especially as compared to legacy options like gas-fired cogeneration. Stanford has the resources to make that investment, but what about a less affluent university or city? With its balance sheet, credit rating, and borrowing capacity, Stanford can afford to think in present value terms. Entities without those advantages, by contrast, know that both capital costs and energy costs are in fact paid in real time. These are some of the next and urgent frontiers for innovation at Stanford. And with the unprecedented resilience risk of this age of climate change going forward, we've seen all of those except earthquakes (knock on wood) in the very recent past.

Next Step Going Forward
Needless to say, we cannot rest on our laurels. What are the next steps? How do we make Stanford's energy system even better, both operationally and in terms of policy outcomes?

There are two interdependent ways to approach that question, making each component better, call that micro to macro, and making the entire system better, macro to micro. It's important to note that the Stanford system is particularly well positioned for the macro- to micro-optimization. Because we are the utility but also our own customers. In fact, one of the groups within SUI is Facilities Energy Management. That's a big advantage: a typical utility's visibility and control stops at the customer's electric or thermal meter. Because we are all one Stanford, together with our colleagues in each building we see, control, and optimize the entire value chain from the solar farms to the electrical outlet and air handler. At the same time, we need to optimize and position the university to thrive as our policy goals and operational needs around energy and its myriad uses might change.

What does that mean? What is the future for Stanford's energy system—SESI X.0, so to speak?

- Inbound transmission and distribution system hardening
- Electrification
- The last 20% of decarbonization from
  - Diesel-powered emergency generators
  - Petrol-powered vehicles
  - Natural gas-powered clothes dryers and lab burners
  - More energy storage
  - Movement toward 24-7 clean energy
  - Opportunities to make our buildings and our entire campus smarter and more efficient

To make this real, we are working closely with our faculty, student, and staff partners on a number of key initiatives:

- Lake Water Heat Exchange, providing an additional heat sink to maximize chilling capacity on hot days
- Building Decarbonization, finding that last 20% or so of GHG emissions
- Room-Level Thermal Forecasting and Control, to better optimize system performance and thus "stretch" existing assets across more buildings, greater electrification, and more frequent and severe heat storms

Stanford 2025 Climate Action Plan (CAP)
Finally, our new—and newly comprehensive—*2025 Climate Action Plan*, which is currently underway.[10] The new CAP:

- Goes beyond energy and GHG to examine water use, waste management, transportation access and decarbonization, food systems, reducing scope 3 emissions of all kinds, and circular and nature-based solutions, among other topics
- Goes beyond mere mitigation of climate change to adaptation, resilience, environmental justice (such as local air pollution), and climate justice (like risk from rising sea levels and wildfire smoke)
- Goes beyond merely changing the ways we do things to doing less with more
- Is designed as a dynamic, ongoing process with a robust governance structure rather than a static, "snapshot" plan

Each of these efforts are exemplars of our "Stanford as Living Lab for Sustainability Innovation" ethos. We host hundreds of tours of the CEF every year, including students, researchers, and industry from around the world. Within

---

[10] https://sustainable.stanford.edu/operations/energy-climate/planning

operational limits, we provide data, component, and system access across the energy system to collaborative research efforts, like the room-level thermal forecasting and control project mentioned above. We have also inaugurated a *Living Lab Fellowship for Sustainability*[11] together with Stanford's new Doerr School of Sustainability[12] and Stanford's Bill Lane Center for the American West.[13] Fellows are drawn every year from both the Stanford undergraduate and graduate student ranks to solve real-world operational sustainability challenges at Stanford, together with mentors and advisors from both operational staff and academia. All of these efforts reflect the same spirit as SESI itself: optimizing for Stanford and providing a template for the rest of the world.

In the meantime, we monitor our progress with sophisticated (but always improvable) instrumentation and control within the system as well as detailed cost, carbon, and renewable energy accounting for the system as a whole. These results are reported extensively within the university and, with respect to carbon and renewable energy, to the State of California.

Again, operations and policy. That's an extraordinary scope, depth, and complexity of optimization and positioning, both at the component and the system level.

This, in turn, begs the question of how artificial intelligence can help us get the most out of the energy, water, real estate, carbon, and capital that we use to make sure that teaching, research, and healthcare are the best that they can be at Stanford, now and into the future.

Summary

Stanford University is committed to making sustainability innovation real: to sustain life on earth, to accelerate solutions for humanity, to catalyze discovery in every field, and to prepare citizens and leaders for lives of active citizenship. Stanford is committed to ensuring equity and inclusion in our research and on our campus, embedding ethics across research and education, and engaging with partners beyond our walls to learn from and give back to our local and global community.

---

[11] https://sustainable.stanford.edu/living-lab/fellowships

[12] https://sustainability.stanford.edu/

[13] https://west.stanford.edu/

In addition to teaching, research, and healthcare, Stanford is a self-contained, full-service, and vertically integrated city. As such, it operates its own mission-essential electricity and thermal energy system, water and sewer, waste manage-ment, civil infrastructure such as roads and bridges, and building automation and energy management systems.

Within that Stanford's climate action has historically focused on energy and, within the energy space, on scope 1 and 2 greenhouse gas emissions. So far, we've been remarkably successful in that effort, reducing those emissions by a remarkable 78% from their peak with the SESI project, but that is only the beginning. Stanford's ultimate GHG goal is now net-zero scopes 1, 2, and 3 by 2050, and is "walking the walk" on other sustainability topics, integrating innovations in decarbonization, water conservation, and waste reduction, reuse, recycling, and composting into the very fiber of its campus.

Stanford's climate action planning currently underway remains focused on energy and GHG but, for the first time (and going beyond the confines of its 2050 goal), (1) includes water, waste, food systems, transportation, and the investments made by its endowment, and scope 3 topics such as supply chain, travel, the built environment, and employee commutation; (2) works across Stanford's decentraliza-tion, bringing together academia, students, alumni, operations, our endowment, out-side stakeholders, and partners in higher education, industry, and the Silicon Valley and global startup communities; and (3) expressly includes climate adaptation, resilience, and environmental and climate justice considerations.

More broadly, Stanford seeks to be a "living lab" for sustainability, with academ-ics and operations working together to use our campus systems as laboratories for research and technology application. In doing so, we must carefully balance the mission-essential nature of our systems with opportunities to derisk new technolo-gies, business practices, and change management processes.

Making sustainability innovation real is an enormous challenge but also an enor-mous opportunity. In that spirit, we are working every day to optimize outcomes for Stanford and provide templates for the rest of the world.

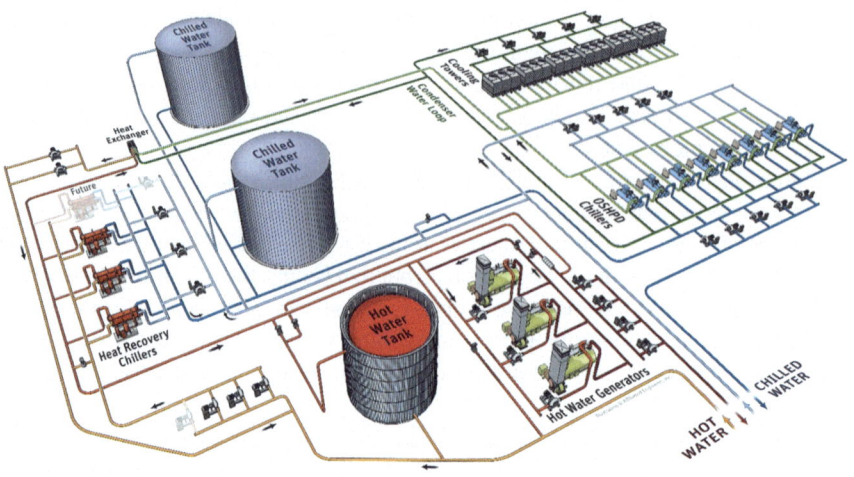

Authors (Gitte Haar) comment on Stanford

- Important showcase with proven, implemented, and science-based methodology to inspire and follow,
- Financials and predictions on future risks, costs, prices, and asset valuation have been an important part of the overall decision scheme, which is often overseen by many today. The cost/price levels of today are often seen as normal not realizing the financial effects of doing nothing and that transition renewable and adapted systems are part of securing the future assets and credit valuation as well as ensuring stable and predictable operation.
- Monitoring is emphasized as an important element in the new Stanford 2025 Climate Action Plan. As well as widening the scope to other resources than energy and to scope 3.
- Showcase on how important it is to have a close collab between academia and implementation to drive the transition to a sustainable future, and the history for this seems stronger in USA than in Europe.
- Systematic sharing of knowledge and methods and taking a forefront position on the transition is very important to spread the work at universities to society.

4.2   City of Amsterdam

4.2.1   Implementation Agenda for a Circular Amsterdam 2023–2026

By: **Zita Pels**, Deputy Mayor for Sustainability and Circular Economy & **Sofyan Mbarki**, Deputy Mayor for Economic Affairs

Introduction

Sustainability is no longer just an aspiration; it is a dire necessity. In many countries around the world, climate change is already a matter of life and death—a humanitarian disaster that often affects those who have contributed least to its causes. The challenges go further than the climate.

Here in the Netherlands, by April we have already consumed the resources with which we should make do for an entire year. In other words, if everyone on earth lived like a Dutch person, we would consume 3.6 earths this year, according to calculations by the Global Footprint Network.

More than 80% of Amsterdam's CO_2 emissions are caused by the use of materials. This CO_2 is not necessarily emitted in our city, but it *is* emitted by the people of Amsterdam. This means we have a major impact on communities beyond our borders. Because of our meat consumption, rainforests elsewhere are felled to produce cattle feed. We see more and more images of growing piles of discarded textiles and electronics in countries such as Ghana or Chile. And some of the plastic waste we produce eventually ends up in the oceans.

The economy has to be organized differently. The UN report Turning off the Tap (May 2023) shows that much is already possible. Using existing techniques, global plastic pollution can be reduced by 80% by 2040.

Redesigning our economy is a great opportunity to make it greener and fairer. We can do this by helping the most vulnerable people in our city and by taking responsibility for what happens in the countries where our goods and raw materials come from.

That is why we facilitate textile companies which reduce the amount of water needed to make jeans, for example, or turn old clothes into new yarns. We will make it easier and cheaper to have appliances repaired so that fewer new raw materials need to be extracted. We will provide more facilities to separate organic waste so that less peat from vulnerable areas is needed to produce compost. Where linear incentives are still present in economic policy, we will phase them out.

The transition to a circular economy is essential and not optional. We have no more time to delay difficult measures. We need to accelerate the pace of implementation. The Implementation Agenda for a Circular Amsterdam 2023–2026 sets out an overview of what the City itself does and what we make available, so that as many parties as possible can contribute to the transition to a circular economy. This transition will become increasingly visible in the city in the coming years.

Doing what we can do now

The road to a circular economy will be long and demanding, and the transition will sometimes be a source of discord. We will need to overcome great difficulty and resistance along the way. We also don't know everything yet; experts still need to work on the necessary research and studies. But this should not be an excuse to delay action. We are not waiting for new laws and regulations from central government or the European Union. We will do what we can do now. And at the same time, we will keep making proposals to the government for stronger measures.

In times of transition, the government must provide clarity about the future. The more clarity, the safer the investment climate. As circular norms and standards become more specific in the coming years, Amsterdam will continue to make the transition. With this implementation agenda, we aim to send a strong signal that the City is an ally for all Amsterdammers who want to help shape the circular economy.

The time for talking about it or waiting for others to act is over. Only by working together will we succeed. By acting, we inspire each other and that generates energy—energy to ensure that the generations to come have a future.

Zita Pels-Deputy Mayor for Sustainability and Circular Economy

Sofyan Mbarki-Deputy Mayor for Economic Affairs

Implementation Agenda for a Circular Amsterdam 2023–2026 has four chapters and an appendix.

Ch1: State of Play, which includes the goals, concepts, and status around the use of materials

Ch2: Together with the City, which explains what the municipality will do to further enable entrepreneurs and social initiatives to contribute to the circular economy

Ch3: Actions per Value Chain, which explains what steps are being taken for each of the three value chains (food and organic waste streams, consumer goods, and the built environment), including what the City is doing to reduce its own environmental impact

Ch4: Cross-cutting themes, which describes actions being taken by the municipal executive to remove barriers to the transition to a circular economy

Appendix: Outline of the proposed budget for the funding provided for in the coalition agreement (Amsterdam Agreement, June 1, 2022)

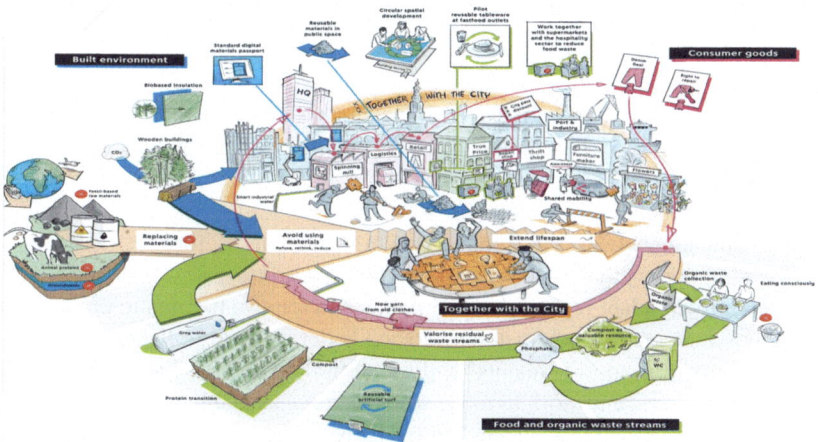

Executive Summary 🎲

In a fully circular economy, materials and products are used and reused almost endlessly. This means we avoid waste and pollution. Renewable raw materials, such as wood or flax, play a key role. In a circular economy, we will live within planetary boundaries, keeping the social foundation in mind. We will then no longer be dependent on new, non-renewable raw materials, such as oil and gas, which can run out or may come from geopolitically unstable regions.

Achieving a circular economy is not an end in itself. It is a way of working and living that produces a climate-neutral, equitable economy, within the planet's limits. A circular economy will:

- Reduce greenhouse gas emissions
- Reduce human impact on biodiversity
- Improve the quality of the living environment
- Improve the security of supply of raw materials

Currently, we are still far from a fully circular economy, but we are making strides. The City of Amsterdam wants to help its residents and entrepreneurs take responsibility in the transition to a circular economy. The City plays a role in bringing together stakeholders, creating opportunities and removing barriers.

We realize that the City must act as a role model. In our own procurement, in the maintenance of municipal buildings and public spaces, and in issuing tenders, circularity is increasingly our guiding principle.

Measures which the City will implement include the following:

(A) Together with the City

- Amsterdam is going to make it easier for businesses in the city and in the Port of Amsterdam to operate according to circular principles. Arrangements will be put in place for:
  - Individual support.
  - Strengthening networks and connecting organizations.
  - Providing more physical and environmental space.
- Amsterdam will adapt regulations and policies to remove barriers to circularity where possible, including adjustments to and better support for licensing.
- Amsterdam will support residents' and social initiatives, with knowledge, contacts and resources, and work to make suitable space available.
- Amsterdam will organize a citizens' council on the topic of waste in 2024, where residents and business owners will discuss the city's waste challenges with each other, experts, and officials. The citizens' council will formulate proposals that will be submitted to the City Council by the municipal executive.

(B) Food and organic waste streams

- Amsterdam will draw up a food waste action plan in 2024 to ensure that less food is wasted and more is used by social initiatives such as food banks.
- Amsterdam will ensure that more organic waste is used for compost or the production of green gas.
- Amsterdam will support initiatives by the Water Authority Amstel, Gooi en Vecht (AGV), for example, to extract more nutrients from wastewater and return them to the chain.

(C) Consumer goods

- Amsterdam will ensure that there is more space and opportunity for the sharing economy, repairs, second-hand sales, and rentals.
- Amsterdam will prioritize strengthening the network of circular textile companies.
- Amsterdam is expanding the City Pass discount for clothing repair to include, at the least, the repair of electrical appliances.
- Amsterdam is looking for businesses that would like to experiment with reusable tableware instead of single-use tableware.
- Amsterdam itself will use fewer consumer goods, more frequently opt to repair rather than replace, and purchase circularly.

(D) Built environment

- Amsterdam will help homeowners insulate and renovate according to circular principles.
- Amsterdam encourages the use of bio-based building materials to reduce the harmful impact of the production of building materials.
- Amsterdam is setting a good example by entering into a framework contract for the circular procurement of 30 new school buildings.
- Amsterdam reuses existing materials for the design of public spaces, unless there is no alternative. Artificial turf fields, for example, are tendered on a circular basis.

(E) Establishing preconditions

- Together with other governments, universities, and research institutions, Amsterdam is establishing the preconditions for the necessary system change. We are contributing to unambiguous national standards and definitions, applying material passports, monitoring material flows in the city, and working to adjust regulations that are still hindering circular initiatives, such as the application of true pricing in investment decisions and procurement, and certain licensing criteria.

The measures outlined above are, by definition, incomplete. There are many parties both within and beyond our city that are already contributing to the transition to a circular economy. At the same time, no one can yet envisage a detailed picture of what needs to be done to realize a circular economy. We expect that within the period of this implementation agenda we will gain insights into how the roadmap to a circular city will look.

Municipal elections in 2026 will see the start of a new 4-year administrative term. By this time, based on the results and lessons learned from this implementation agenda, progress in the **National Circular Economy Program (NPCE)**, and new European policies that are being prepared, it will be possible to present a recommendation to the City Council on how to achieve the 2030 and 2050 goals.

4.2.2 Ch1: State of Play

Amsterdam is regarded worldwide as a pioneer in the transition to a circular economy. From the first study of the circular economy in 2015, the City has been working with national and international partners based on the ideas of the iconic doughnut model, with inclusive prosperity as the goal. The social foundation is the inner ring of the Doughnut, and its outer ring is formed by the planetary boundaries. The goal is to thrive in a space within those rings which is both ecologically safe and socially just. Amsterdam is working to reduce the impact of material use while considering social and ecological values.

On May 19, 2020, the City Council adopted the Amsterdam Circular Strategy 2020–2025, which included the Innovation and Implementation Program for a Circular Economy 2020–2021. On May 24, 2022, the City Council was informed in writing of the lessons and recommendations drawn from this program. Based on this report and the funds made available in the Amsterdam Agreement of June 1, 2022, the Implementation Agenda for a Circular Amsterdam 2023–2026 was drafted.

In this administrative term, the emphasis lies on implementing the direction set out in the strategy and the implementation agenda. Decisions about a possible adjustment of this direction will be prepared for a subsequent administration. Therefore, the strategy adopted will not end in 2025 but, together with the implementation agenda, will remain in force throughout 2026.

The City of Amsterdam cannot realize the transition to a circular economy alone. We are shaping it together with partners, including entrepreneurs, residents, social initiatives, universities and research institutions, water authorities, other municipalities, provincial government, central government, and the European Union.

On March 17, 2023, central government published the National Circular Economy Programme (NPCE), which includes the goals that Amsterdam also uses. The NPCE can be seen as the first step in the leading role that central government will take to ensure coherence and direction during the acceleration of the transition to a circular economy.

Fortunately, we do not have to devise the necessary measures alone. We are building on the support offered by the province of North Holland, central government, and the EU. We are learning by doing. We are learning together with Milan,

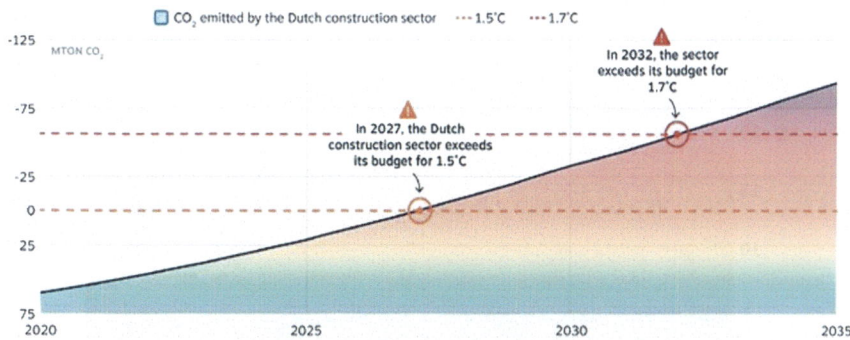

Fig. 4.1 CO_2 **emission by the Dutch Construction Sector**

London, Lille, Rotterdam, Amersfoort, Haarlem, Gelderland, and many other administrations. In addition to problems with nitrogen levels in the Netherlands, we are also approaching the CO_2 ceiling (Fig. 4.1).

Development of Material Use

The latest edition of the Amsterdam Circular Monitor (March 27, 2023) shows a worrying picture: material use is not decreasing in Amsterdam. This development can be seen worldwide and is in line with the ICER report by the Netherlands Environmental Assessment Agency (PBL) of January 26, 2023. *Circle Economy* writes in its latest Circularity Gap Report that the global economy has extracted more resources from the earth in the past 6 years than in the entire twentieth century. The impact of material use is also far greater than previously thought. On the next page the key insights from the Amsterdam Circular Monitor.

1. **Direct CO_2 emissions have been calculated** for the city of Amsterdam for some time. "Direct" emissions are those caused using energy for heating, transport, and electricity in Amsterdam. The energy transition focuses primarily on reducing emissions in Amsterdam, yet huge sustainability gains (climate, biodiversity, living environment) can be achieved by using fewer primary raw materials elsewhere in the chain and thus reducing CO_2 emissions outside Amsterdam, also known as "indirect" emissions. A mobile phone, for example, requires rare metals to be mined, petroleum to be extracted for the manufacture of its plastic components, and energy to be used for transport and manufacturing before it reaches the stores here. The latest monitor included these indirect emissions for all the city's material use for the first time. Figure 4.2 shows that the CO_2 emissions caused elsewhere by Amsterdam's material use are four times greater than the city's own CO_2 emissions.

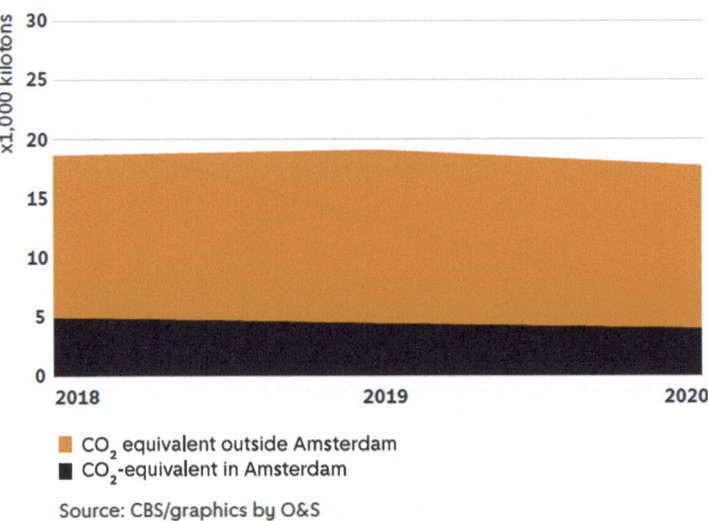

Fig. 4.2 CO_2 emissions within and outside Amsterdam because of material use in Amsterdam

Environmental costs of material use

Amsterdam uses the Environmental Cost Indicator (ECI) to manage the negative environmental impacts of extraction, production, and transportation of
products. The ECI adds up the cost of preventing our material use from causing negative environmental impacts. These preventive costs are currently not
paid by the producer or included in the price for the end user. As a result,
environmental damage must be repaired after the fact. In most cases, this will
result in higher costs than if the environmental damage had been prevented
beforehand. The consequences of this are beginning to become visible in, for

example, the national nitrogen emissions crisis; the costs are passed on to society and future generations after the fact. The ECI summarizes more than 10 environmental impacts, such as soil, water, and air pollution, climate change, and biodiversity loss, and is expressed in euros. The lower this monetary value, the more environmentally friendly the material. ECI is used as a quality criterion in tenders in the Netherlands by several agencies, including the Directorate-General for Public Works and Water Management.

2. **The use of raw materials** is not decreasing rapidly enough to reach the goal of a 50% reduction by 2030. Fig. 4.3. If this does not change, we will not meet the national goals.

Carbon budget

If indirect CO_2 emissions from material use are included, Amsterdam's carbon budget to keep global warming below 1.5 °C already ran out last year. In other words, if everyone on earth were to live like the average Amsterdam resident, the earth would warm by more than 1.5 °C.

3. **Over 80% of Amsterdam's CO_2 emissions** (from residents, municipality, and companies) **are due to the city's use of materials.** These CO_2 emissions occur mainly in the extraction of raw materials and the production of goods elsewhere for consumption in Amsterdam. The rest of the CO_2 emissions are caused by the city's own energy consumption. To deal with these local CO_2 emissions, the City has drawn up the New Amsterdam Climate: Amsterdam Climate Neutral Roadmap 2050 (3 March 2020) which clarifies how the energy transition in Amsterdam is taking shape. Progress on this roadmap is described in a letter to the City Council dated 30 May 2023.

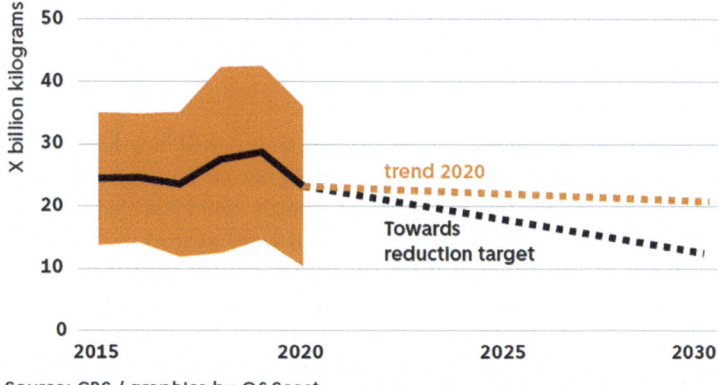

Source: CBS / graphics by O&Scost

Fig. 4.3 Raw material use in Amsterdam

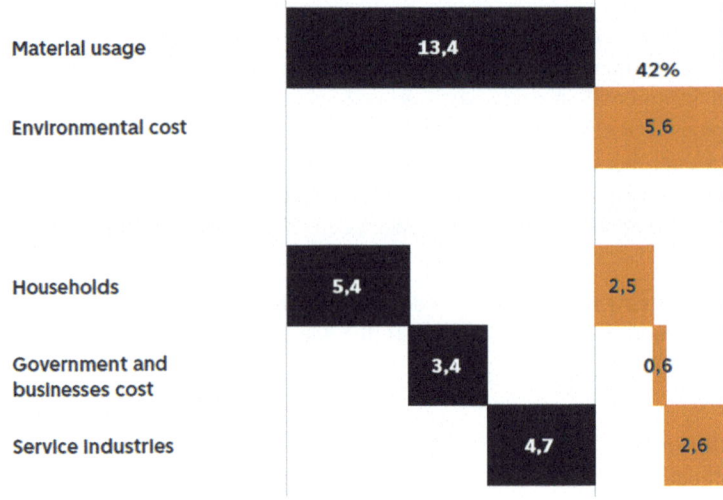

Source: CBS / graphics by O&S

Fig. 4.4 Expenditure and environmental costs of material use in Amsterdam by type of environmental impact, 2020

4. **The cost of preventing the ecological damage caused by our use of materials was estimated at €5.6 billion in 2020.** If these costs were passed on, everything would be 42% more expensive on average. These costs are not currently included in prices and the damage is passed on to society and future generations. In practice, repairing environmental damage is more expensive than preventing it (Fig. 4.4).
5. **The environmental impact of different value chains** is shown in Fig. 4.5. Food and consumer goods have the greatest impact. Food has about the same impact as all other material use combined. Construction uses large quantities of material but has a relatively limited impact per kilogram. We also see that fossil fuels—especially in the form of petrol, diesel, and kerosene—account for a significant share.

Phases in the transition

The transition from a linear to a circular economy will not happen by itself. Figure 4.6 makes it clear that two paths must be followed. Through experiments and pilots, novel solutions must be devised, which must then be scaled to a "new normal." At the same time, old habits, attitudes, and ways of working must be broken down. In relationship to the latter, the diagram includes words such as "destabilization" and "chaos." These elements in the transition will cause discord and bring discomfort and struggle. It is up to the government to prepare for the next phase to emerge, in order to be able to inspire trust and provide direction to get through the more chaotic phases of the transition as smoothly as possible.

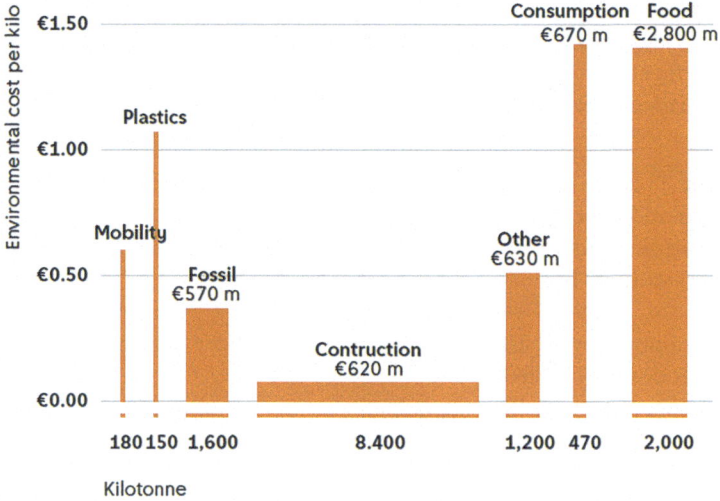

Source: O&S

Fig. 4.5 Expenditure and environmental costs of material use by type of environmental impact

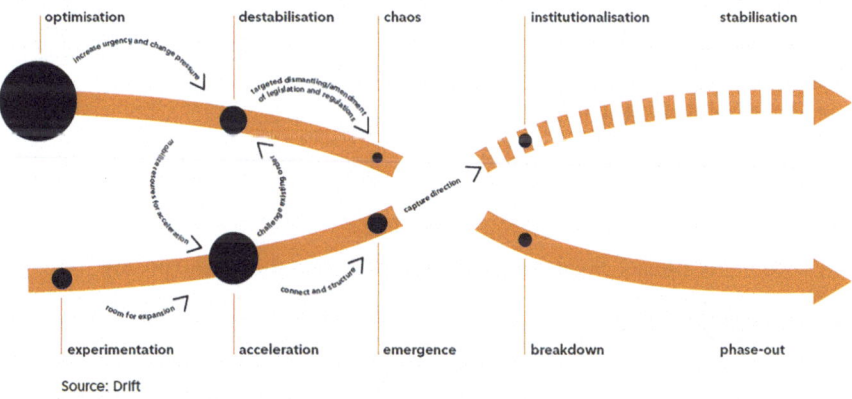

Source: Drift

Fig. 4.6 The DRIFT X-curve, phases in the transition

The phase a transition is in determines the approach: experimentation and research in the early phases, and scaling, securing, and transferring during further development.

Transitions are lengthy and often unpredictable processes. It is essential to speed up and provide direction in these processes. It is also important to continue to identify new as yet unknown developments, provide room for experiments, and ensure that the transition continues to accelerate. Amsterdam is active in all phases of the X-curve, from start to finish, so that we can achieve our goals for 2030, 2040, and 2050.

Four Strategies

The NPCE identifies four strategies to achieve the transition to a circular economy. This implementation agenda deploys all these strategies across the three value chains.

Four strategies for the transition to a circular economy

Prevent

(Design and procurement phase)
Strategies and activities aimed at preventing or reducing material use.
For example, by:

- consuming fewer products or using them for longer (refuse and reduce)
- smart design (rethinking)
- sharing, renting, or leasing products

Extend

(Use phase)
Strategies and activities aimed at lifecycle extension.
For example, by:

- reusing materials (reuse)
- repairing products(repair)
- refurbishing products(refurbish)

Valorize

(End-of-life phase) Strategies and activities aimed at repurposing, recycling, or energy recovery.
For example, by:

- recycling materials for other (lower-value) uses or products (repurpose, recycle)
- recovering energy from waste incineration or waste heat (recover)

Substitute

(Design and procurement phase)
Substituting fossil materials with biomaterials.
For example, by:

- building with wood
- insulating with natural materials (flax, bulrush)
- plant-based food

Goals

We want Amsterdam to be a thriving and equitable city. We want to guarantee a good life for all the city's residents within the natural boundaries of the planet. We want a city in which the prosperity and well-being of all is central. We therefore look not only at the ecological boundaries that we must not exceed, but also at the minimum social foundation, both local and global, which we must maintain. This thinking is comprised in the Amsterdam City Doughnut model.

By signing the National Raw Materials Agreement in 2017, the City of Amsterdam has made a commitment to the national goals of using 50% fewer new, non-renewable raw materials by 2030 and having a fully circular economy by 2050.

This promise is embedded in the Amsterdam Circular Strategy 2020–2025 (May 19, 2020).

Procurement

- By 2030, the City's purchases will be 100% circular, and the City will have reduced its own consumption by 20%, starting with consumable materials, and the fixtures and fittings of our own premises and, where possible, of civic buildings.

Food and Organic Waste Streams

- By 2030, 75% of all households will have their organic waste collected separately.

Built Environment

- As of 2022, all new designs for regional developments (including conversions and re-purposing) and public space in Amsterdam are based on circular criteria. We act in collaboration with Amsterdammers, the market, and other authorities. The challenge is to collectively reduce the use of primary raw materials in the built environment to achieve the 2030 target. To this end, we are deploying instruments at city level, such as thematic studies, and at regional level we are working out specific, achievable goals and instruments.
- From 2023, the City of Amsterdam will apply circular and social criteria when working on buildings and in public spaces. This includes procurement, tendering, and tenders for land allocation, and applies to all life phases, from new construction to property management to the end of functional life, unless this is not possible in its entirety.
- From 2025, 50% of renovations and property management in Amsterdam will be carried out according to circular principles. This will include existing social and private housing, civic real estate, schools, utility buildings, and public spaces (both above and below ground).

The transition to a circular economy will result in a new way of working. This will contribute to the achievement of various social goals meaning they do not need to be added as additional policy objectives (Fig. 4.7).

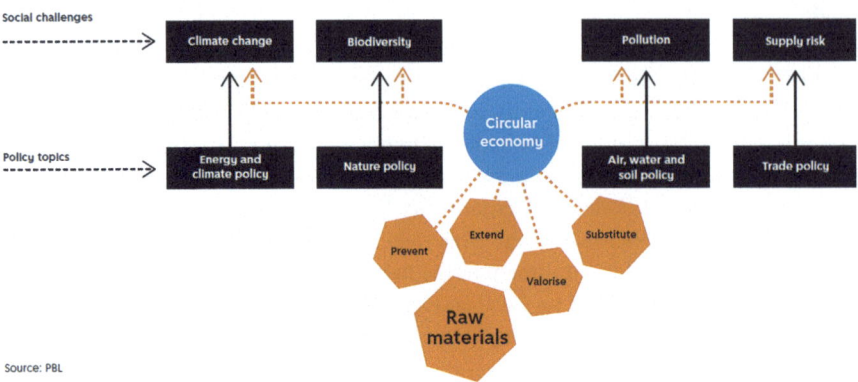

Source: PBL

Fig. 4.7   Connections between the various social challenges

Conceptual framework

Many different concepts and goals are used in relation to the circular economy, such as inclusive prosperity, doughnut economy, community well-being, and the UN Sustainable Development Goals. These concepts demonstrate our aim to consider the indirect social and environmental consequences of our actions when making decisions. Figure 4.8 shows the key elements for the context of Amsterdam.

By the end of 2023, via a more detailed development of the concept of "inclusive prosperity," the municipal executive will consider the differences and similarities between the various concepts in greater detail.

See: https://statistiek.amsterdam.nl/publicatie/de-staat-van-de-stad-amsterdam-xi-2020-2021

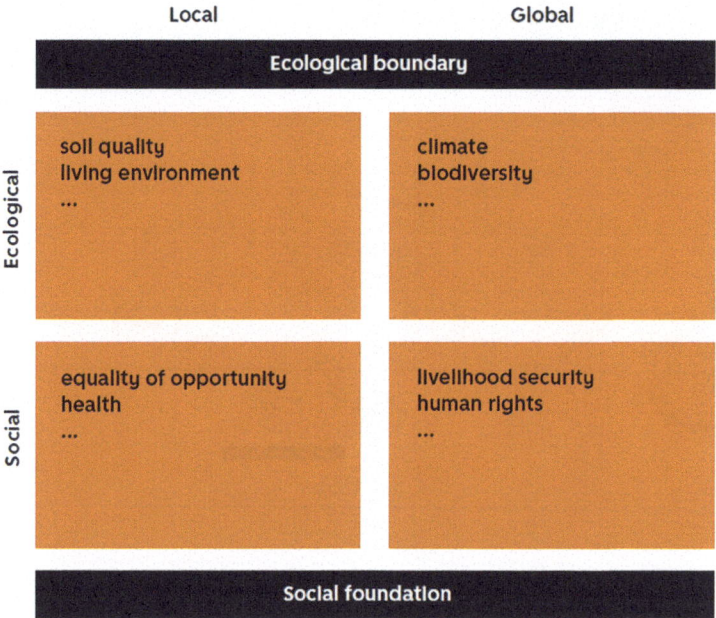

Fig. 4.8 **Selection of key social impacts within the doughnut economy**

A circular economy aims at:

- Reduction of greenhouse gas emissions
- Reduction of human impact on biodiversity
- Improved quality of the living environment
- Improved security of supply of raw materials

The war in Ukraine highlights once again how dependent we are on other countries for the raw materials that are the basis of the products and services we use here. In the period 2023–2025 we are preparing new targets so that these can be presented to the City Council at the start of the new administration period (2026–2030).

4.2.3 Ch2: Together with the City

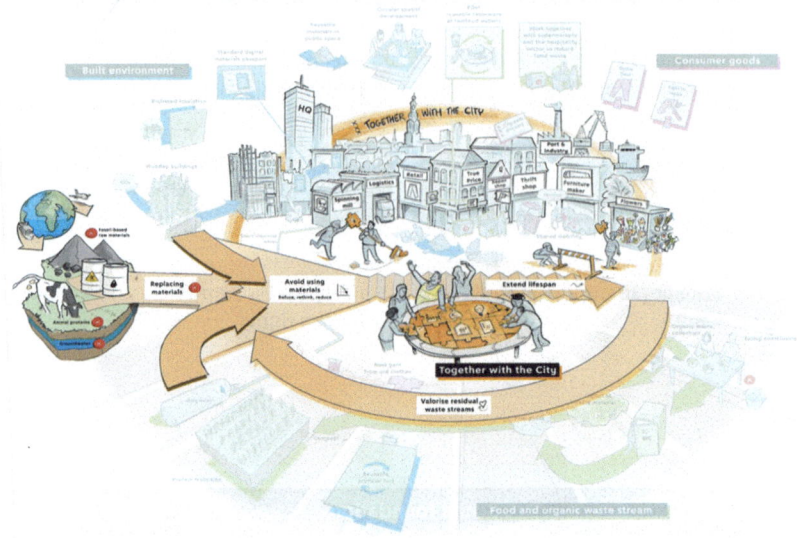

The transition to a circular economy calls for leaders and ambassadors. These people can often be found in small and medium enterprises (SMEs) and social initiatives. The City will support them by removing barriers and offering opportunities.

Ultimately, everyone must join the transition. Many companies and institutions realize this and want to do so, but do not yet know how. The City also wants to offer this group support and a frame of reference for action.

Together with Education and Science

The connection with education and science is of great importance both for the development of knowledge needed globally and for the translation of theory into practice. The City partners with vocational training institutions, colleges, and universities, as well as with institutions such as the Central Bureau of Statistics (CBS), PBL, Circollab, Cirkelstad, AMS Institute, Impact Hub, and CIRCO. We create learning modules as part of the City's support for primary and secondary schools. For example, we offer teaching packages on waste separation in areas in which we will place new collection containers for organic waste. We also periodically organize collection campaigns with schools for textiles or e-waste (electrical appliances and electronic devices, of which some 17,840 were collected by 2022).

The City also contributes to the Circular Workspace, where students, teachers, and researchers work to make SMEs more circular. Questions from SMEs are the starting point. The initiative brings value in two directions: courses at educational institutions benefit from working on concrete cases from the business world, while SMEs are encouraged to get started with circularity.

In this way, we ensure that circularity is embedded in a new generation of professionals.

Using robots to reuse building materials

The City supported the Circular Wood for the Neighbourhood project. This involved experts from the Digital Production Research Group of the Amsterdam University of Applied Sciences using robots in housing corporations' renovation projects to make the reuse of wood easier and more efficient. Robots and parametric design were used for purposes such as making furniture. This project shows how the innovative capacity of education helps in the transition to a circular economy.

Creating Circular Jobs

The share of circular jobs in the Amsterdam labor market grew from 5.7% to 6.5% between 2011 and 2021: an absolute increase of over 16,000 jobs.[2] And while more jobs mean more social opportunities for Amsterdammers, this is not all good news; labor shortages are a threat to the transition to a circular economy. Alongside the growth of circular employment, there will also be a decline in jobs, for example in

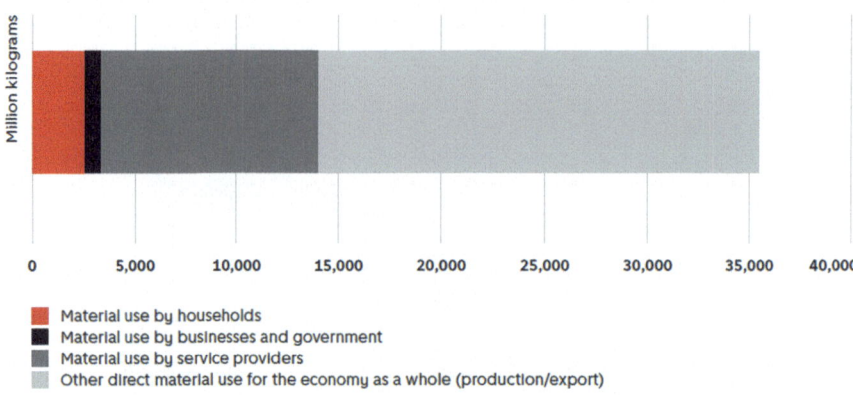

Fig. 4.9   Direct use of raw materials across the economy divided by use

the fossil sector. City policy states that coal transshipment will disappear by 2030 at the latest. This can be perceived as a threat by workers. To provide solutions, we are bringing entrepreneurs and job seekers together in the Job with a Future program (Fig. 4.9). https://onderzoek.amsterdam.nl/publicatie/circulaire-werkgelegenheid-amsterdam

We also support entrepreneurs in finding, training, and coaching workers who want to make the transition from another industry to sectors that can contribute to a circular economy. These include construction, installation engineering, infrastructure, and the green sector, but also less obvious sectors such as procurement and marketing. According to a 2021 study on the employment effects of Amsterdam's circular policy by *Circle Economy*, TNO, and House of Skills, the transition to a circular economy would create an additional 8500 jobs on balance by 2030. Attracting and investing in enough suitable people is therefore even more important.

Business

Enterprises with circular goals, both commercial and non-profit, face many challenges, given that the economy is still primarily linear in structure.

Amsterdam is developing solutions in cooperation with business owners and industry associations to address these challenges and provide solutions to various barriers and bottlenecks.

Challenges include:

- Unfamiliarity with legal, tax, and technical solutions
- Insufficient connection to a relevant network of circular companies
- Insufficient organization of the value chain (suppliers, logistics providers, retail, etc.)
- Lack of affordable physical space
- Uneven playing field; conventional companies do not include many shadow costs in their price making it harder for circular companies to compete
- Insufficient power due to grid congestion, which will limit the establishment or expansion of companies in Amsterdam in the coming years

In response to these challenges, the City is undertaking the following steps:

- Gaining insight into material flows and finding opportunities for action
- Searching for new circular earning models (from products to services, paying for use instead of possession, sharing economy, true price, etc.)
- Stimulating circular business operations
- Shaping collaboration in the value chain (different purchasing, valorizing residual waste streams, agreements with suppliers and customers, etc.)
- Using the City's own purchasing power to act as a launching customer for new, circular solutions through circular tenders
- Ensuring that relevant, practical knowledge of current and future European regulations in the field of circularity is readily available to the business community

Textile innovation platform BYBORRE

BYBORRE provides an online platform where designers can create their own textiles based on sustainable raw materials and then produce them on-demand.

We are going to expand our support for businesses. We will do this in consultation with business owners, sector organizations, and other relevant stakeholders. We will look closely at existing activities and regulations, such as those of the Province of North Holland, to connect with them, strengthen them, or supplement them where necessary. Where the City does not have the authority to make changes, for example regarding creating a level playing field for circular enterprise, we will lobby central government and the European Union. New elements in the support we offer include:

S1 Individual support

Starting in 2024, Amsterdam will hire external capacity to provide tailored advice on adopting more circular business practices to approximately 100 businesses annually. This is a continuation and expansion of an advisory program previously made available to companies in the manufacturing sector. Companies are connected to technical solutions and relevant networks. They also receive help with legal questions, for example, or advice on how to future-proof their business in terms of internal organization (knowledge, support base) and business models.

There are many new European regulations pending that will set new requirements. The Corporate Sustainability Reporting Directive (CSRD) and the requirements around the right to repair, for instance, will oblige more and more companies to report on the impact of their activities on people and the environment. The City will ensure that relevant, practical knowledge is made readily available to businesses.

S2 Support for collective action

Starting in 2024, we will facilitate two collectives of business owners annually in developing frontrunner groups, as was previously done in setting up the Circular Hotels frontrunner group. The City aims to use this tool to achieve more cooperation within the value chain. Specific target areas for the City include:

- Textiles: various initiatives throughout the chain, from the design and production phase and the purchase and use phase to repair, processing, and reuse
- Appliances: primarily focused on life extension, such as repair, second-hand stores, and overhaul, as well as initiatives based on pay-for-use rather than ownership

We will also provide enterprise collectives such as Business Improvement Districts (BIDs) information about how they can avoid single-use plastics, for example, or replace them with sustainable, reusable alternatives.

S3 Stimulating ecosystems for innovation

Many SMEs are already supplying innovative products and services in Amsterdam that contribute to the transition to a circular economy. The City aims to strengthen the climate for innovative enterprise by collaborating with governments, businesses, research and educational institutions, and civil society organizations. Each player in this network makes a unique contribution.

To achieve this, Amsterdam will:

- Create physical and environmental space for circular initiatives

- Actively connect organizations
- Make resources and knowledge available to develop business cases, including alternative ways of including value and environmental costs, such as true pricing
- Make support available, as previously described

We call this approach an "innovation ecosystem." Suitable sectors are sustainable construction, textiles, and food and organic waste streams.

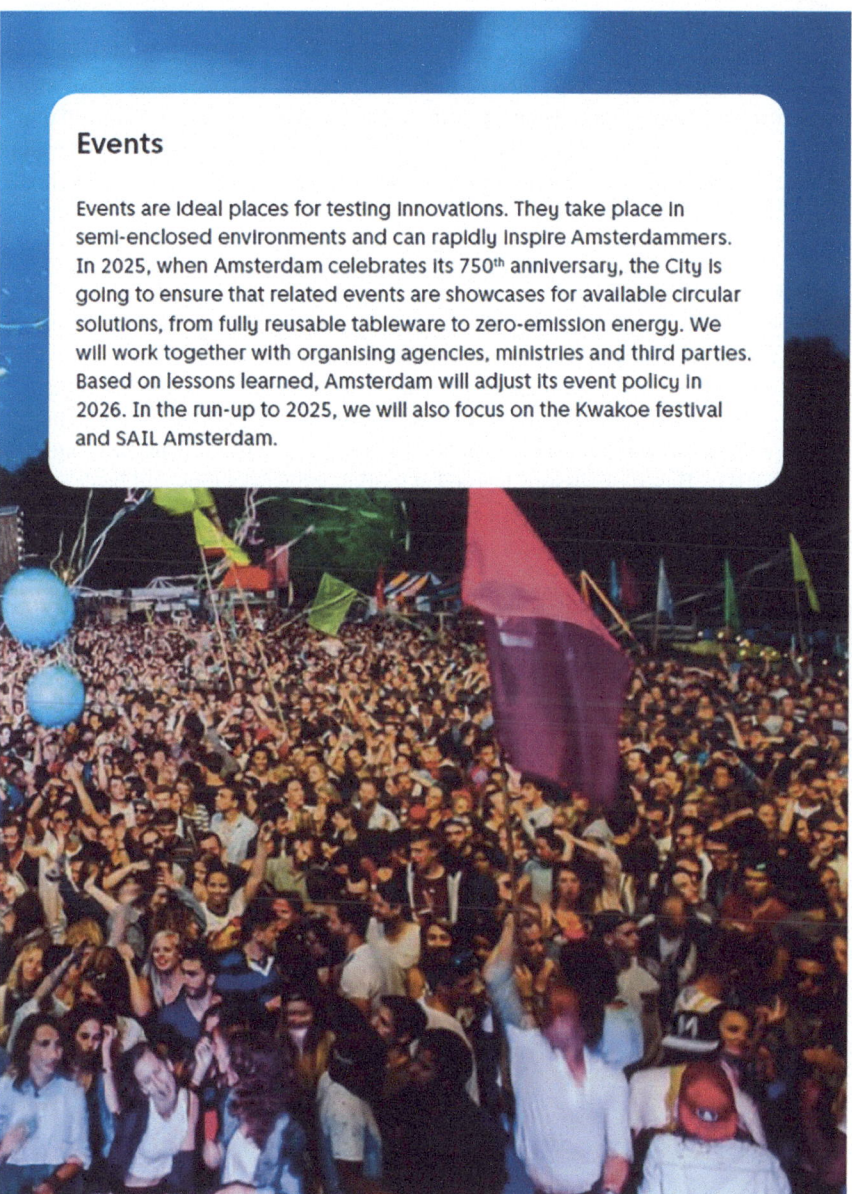

Events

Events are ideal places for testing innovations. They take place in semi-enclosed environments and can rapidly inspire Amsterdammers. In 2025, when Amsterdam celebrates its 750th anniversary, the City is going to ensure that related events are showcases for available circular solutions, from fully reusable tableware to zero-emission energy. We will work together with organising agencies, ministries and third parties. Based on lessons learned, Amsterdam will adjust its event policy in 2026. In the run-up to 2025, we will also focus on the Kwakoe festival and SAIL Amsterdam.

S4  Intensify collaborative learning with entrepreneurs and social initiatives

The City wants to strengthen the feedback loop with enterprises and social initiatives so that we learn more quickly where bottlenecks lie. We can use this knowledge to adjust our policies, or to improve our procurement conditions, enforce regulations more effectively, and provide better support.

S5  Create space for circular initiatives and business activity

The City will ensure that physical space becomes available for target groups such as thrift stores, repair workshops, and sustainable processing businesses by offering them temporary lease contracts on a priority basis, in part to realize the desired retail diversity. To achieve this, where possible, the City will purchase, rent, or develop property, for example in urban regeneration areas. The City Council will be regularly informed about progress.

Plastic-free hospitality

Following the successful 2021 Plastic-free Terraces pilot in the Knowledge Mile Business, a follow-up project has been launched.

The Plastic-free Hospitality project helps business owners apply circular principles such as avoiding single-use plastics wherever possible or replacing them with sustainable, reusable alternatives. The project is being implemented in two BIDs. Based on lessons learned, a toolkit and workshops will be developed in order to scale to enterprise collectives in the rest of the city.

Green hotels

In recent years, a group of leading hotels has worked with the City to make the catering and use of linen, energy and water for Amsterdam's 88,000-plus hotel beds more sustainable. By 2022, the collaboration between the hotels had grown to become an independent foundation, Green Hotel Club. The foundation aims to encourage more and more hotels and related businesses to join, enabling them to make an even more impactful contribution to the City's circular goals, and to future-proof the hotel industry.

Port and industry

The port area is key to realising the circular ambitions of the city and region. The presence of port logistics, the possibilities for storage and transhipment, and the concentration of raw materials and residual waste streams combine to make the port a place where it is possible to create space and scale for innovation and industrial activity. The port creates more opportunities for high-quality processing of waste streams by facilitating companies working on mechanical and chemical recycling.

The City and Port of Amsterdam aim to further develop the circular industrial ecosystem in the coming years. We are intensifying our cooperation and will realize our circular ambitions in the port via the following two paths:

S6 🌀 Strengthening circular initiatives

- We will continue our cooperation with and support of the Circular Network. This network, established by ORAM and the Port of Amsterdam, brings together companies, universities and research institutions, and authorities with the aim of growing circular industrial activity in the North Sea Canal Area (NZKG).
- We collaborate with the Port of Amsterdam on socially responsible supply chains, building on the authority's efforts to make supply chains transparent and traceable by making agreements with companies that open offices in the port area.
- We help SMEs in the port area with consultancy on circular processes, and with finding financing opportunities.
- We identify opportunities to accelerate circular development of specific chains as a launching customer.
- We promote knowledge transfer and education to bring schoolchildren and students from Amsterdam and the region into contact with the port and circular industry.

S7 🌀 The necessary preconditions and removal of bottlenecks

- With input from the Port of Amsterdam and the business community, we are mapping which laws and regulations are obstacles to the circular transition. We are lobbying higher authorities, such as central government and the European Union, for a level playing field.
- Regional coordination is necessary to ensure that circular industrial activities have sufficient space in the right locations, in terms of physical space, environmental impact, and distance from hazardous activities. We therefore participate in the NZKG NOVEX approach, and the issue of space is explicitly included in the development of the new Environmental Safety Policy and the Port Area Vision.

- Where possible, in collaboration with the NZKG Environmental Service, the City is adapting the licensing process to allow promising circular initiatives to start, accelerate, and scale up. We are also looking at how enforcement can be intensified.
- Realizing the circular economy requires a robust infrastructure for electricity, hydrogen, green gas, CO_2, and synthetic fuels. Grid congestion is currently an acute bottleneck. Space for grid connections will remain limited due to congestion until approximately 2028. In cooperation with the Port of Amsterdam and ORAM, we introducing the importance of the circular economy into the public debate about investments in infrastructure and the prioritization of grid connection and expansion.

S8 Smarter use of industrial water

Water Authority Amstel, Gooi en Vecht and the City of Amsterdam will work together with industry and the Port of Amsterdam to make wastewater (effluent) suitable for industrial applications in 2025. In this way, we will preserve clean and increasingly scarce freshwater as a source of drinking water.

In the coming period, we will clarify which frameworks and direction the City, as shareholders, can give the Port of Amsterdam in this transition and how best to contribute to accelerating the circular transition in the port. It is important that the Port of Amsterdam is able to respond to specific situations and fulfill its role and tasks in a way that does justice to the dynamics of the port area. At the same time, as a shareholder, the City wants to exert long-term influence. Our findings will be used in the reassessment of the Port of Amsterdam's 2025–2030 strategic plan.

Community and Residents' Initiatives

Civil society organizations and motivated Amsterdammers are important ambassadors. They make the transition to a circular economy concrete and visible. They can do so in the street or in debating centers, but also through chance meetings in a thrift store, a self-sufficient community center, or while car sharing. Often these ambassadors are not yet sufficiently appreciated, especially when it comes to their social significance, which is difficult to quantify.

S9 Support for social initiatives and community and residents' actions

In consultation with stakeholders, we will strengthen the support offer for social initiatives and community actions that facilitate the circular economy at a neighborhood level. Social initiatives and community actions have ecological, economic, and social impact on the city. In addition, participants build social contacts. The connection between all these social and ecological aspects will be visualized through the detailed development of the concept of inclusive prosperity, the first results of which will be presented to the City Council in late 2023.

The support includes:

- Financial arrangements that meet the needs of social initiatives and community actions, including the provision of sufficient space, and based on the principle

that the availability of facilities in the city must not deteriorate and must be improved by 2026
- Facilitation of a knowledge network
- Support for owners' associations (VvEs) and private homeowners for circular renovation and insulation
- Collaboration with at least five circular initiatives through which Amsterdammers and local residents can take ownership of a local, circular economy

S10 🔀 Citizens' council

The City will organize a citizens' council on the topic of waste in 2024, in which residents, business owners, experts, and officials will discuss the city's waste challenges. The citizens' council will formulate proposals that will be submitted to the City Council by the municipal executive.

Together with 918,000 Amsterdammers

The circularity of the city is the sum of everyday choices about how we work and live. The involvement of the more than 918,000 residents of our city is therefore indispensable in shaping the transition to a circular Amsterdam. In this implementation agenda, we are making space in the coming years to experience together what an alternative, circular future looks like. We are enabling initiatives by civil society organisations and residents to flourish. We are doing this not only because these initiatives add value, but also to learn where the greatest needs and opportunities lie. We are already discussing the issues on a daily basis in numerous contacts between the municipality and Amsterdammers, and this will intensify through the citizens' council on waste, the innovation initiatives around events, and the development of the support schemes announced in this implementation agenda.

The Amsterdam Doughnut Coalition

The Amsterdam Doughnut Coalition is a network of people and organisations enthusiastic about the doughnut economy model and working together to put it into practice. The Amsterdam Doughnut Coalition increases the visibility of initiatives, organises an annual Doughnut Festival and connects with similar organisations elsewhere in the world. The City of Amsterdam participates in and supports the Amsterdam Doughnut Coalition in partnership with organisations including the Amsterdam University of Applied Sciences and employment agency Olympia. Through its participation, the City learns what transformative actions and support are needed for initiatives to flourish.

4.2.4 Ch3: Actions by Value Chain

The most important value chains on which the City can make an impact are:

- Food and organic waste streams, which cause about half of all ecological impact
- Consumer goods, which have the greatest impact after the food chain
- The built environment, in which the City can exercise considerable influence

These value chains represent a huge volume (Fig. 4.10). They are economically significant for the city and make a strong environmental and climate impact. The

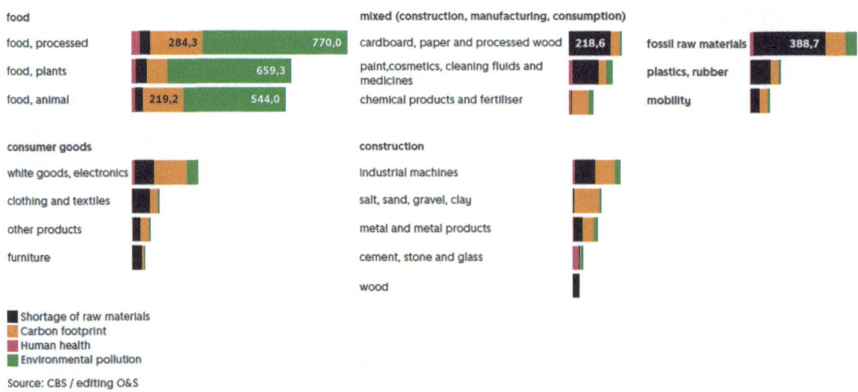

Fig. 4.10 Environmental costs of Amsterdam's material consumption

latest edition of the City's annual Circular Economy Monitor shows that these three value chains have the largest environmental impact. Furthermore, they are areas in which there are opportunities for the City to exercise its influence.

Food and Organic Waste Stream

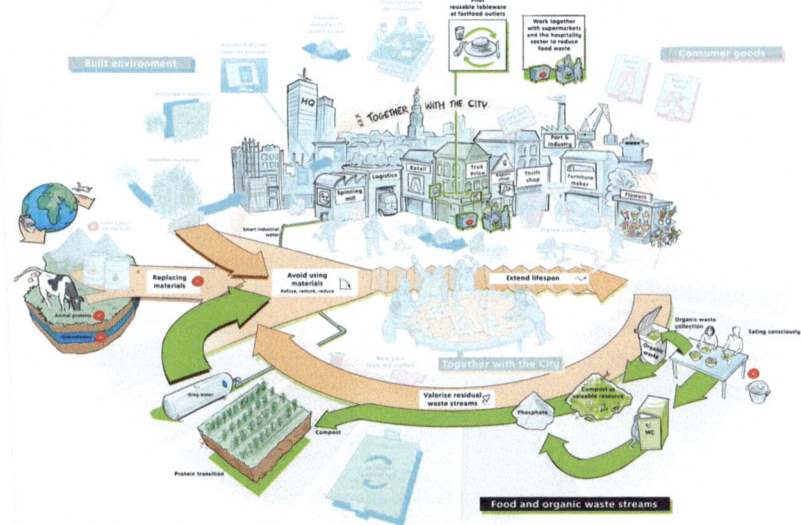

Food and organic waste streams

Our food has an impact (environmental cost indicator, ECI) of €1 billion per year on nature, people, and climate. At almost 20% of the city's total impact, this is the greatest source of damage. There are a number of reasons for this:

- The food system is organized such that raw materials are used only once in linear chains.
- About a third of all food is wasted.
- Organic waste streams are still largely incinerated in Amsterdam, with the loss of valuable nutrients.

This issue calls for a robust intervention. The City's food strategy is therefore being updated in 2023. The Implementation Agenda for a Circular Amsterdam 2023–2026 includes measures that contribute to closing the organic cycle, so we waste less food, and valuable nutrients from the city find their way back to the soil.

V1 📈 Food waste action plan
In 2024, we will draw up the Food Waste Action Plan 2024–2026, together with the National Centre for Nutrition, the Together Against Food Waste foundation, and various stakeholders in the city. We will build on successful examples elsewhere, such as in the province of Gelderland and in Rotterdam.

- Following the example of Milan (winner of the Earthshot Prize 2022), we will "rescue" food that would otherwise be wasted, by connecting more stores, restaurants, and other partners to food banks and social initiatives.
- In consultation with consumers and businesses, we are determining what we can do jointly to prevent food waste. Amsterdam is preparing legal measures to oblige businesses (hospitality, retail, and distribution centers) to combat waste more actively.

Food Strategy Implementation Plan 2023–2026
In 2023, Amsterdam will present an update of the Food Strategy 2023–2026, setting out measures to be taken together with social organizations and educational institutions to widen the available choice in the supply of healthy, fair, and sustainable food for all Amsterdammers. Amsterdam wants to ensure that healthy food is offered in more places in the city, that food waste is prevented, that more local food chains are created, that a more plant-based diet is encouraged and that the City sets a good example by using plant-based foods in its catering as much as possible. The Implementation Agenda includes one part of the program, namely "food and organic waste streams." (Fig. 4.11).

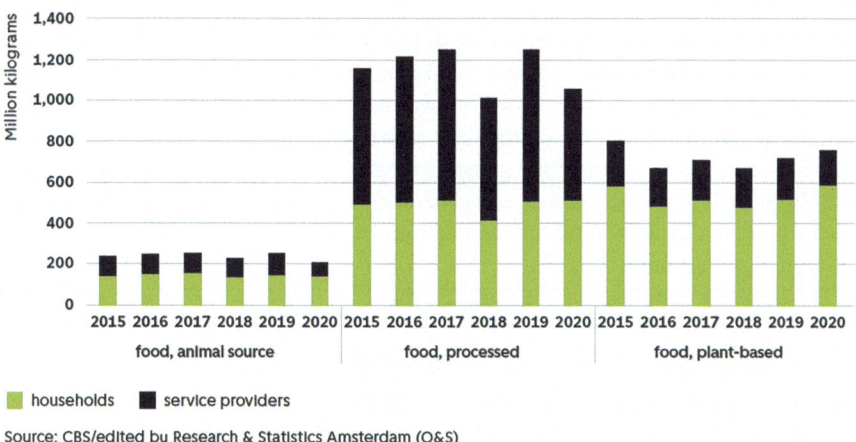

Source: CBS/edited by Research & Statistics Amsterdam (O&S)

Fig. 4.11 Food consumption in Amsterdam 2015–2020

V2 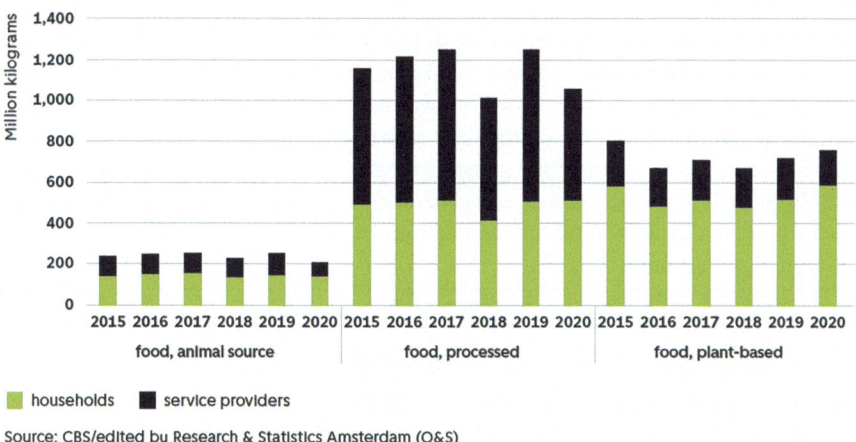 Soil conditioners

Agricultural businesses procure a lot of compost and fertilizers, which requires natural gas and imported phosphate, a rare resource. The environmental impact of this is immense. At the same time, much of the organic residual stream such as sludge or organic waste in Amsterdam is incinerated, because they are impossible to separate from other waste once they have been disposed of in general or household waste. In Amsterdam, around 30% of residual waste is organic.

We will support local and citywide initiatives and businesses that work toward reusing organic residual waste. This can include local composting, waste-to-harvest initiatives, wormeries, and biodigesters but also generating fertilizers from urine from festival toilets.

Because in many cases using urban compost and fertilizers in agriculture is banned, we are researching what is legally possible, in partnership with national organizations that develop new, food-safe processing methods. We also focus on safety and added value—high-quality compost and fertilizers—for the soil and the farmers.

The worldwide importance of circularity in water management is becoming increasingly evident. Shortages of freshwater and groundwater are becoming more frequent, and the nutrients we flush away with the water are becoming scarcer. By finding better ways to reclaim them, we can avoid having to import newly extracted nutrients.

Through Waternet (the water company for Amsterdam), the City of Amsterdam and the Water Authority Amstel, Gooi en Vecht work closely together to make all aspects of water management sustainable: from the production and distribution of drinking water and the collection and treatment of wastewater to the maintenance of canals and dykes. This partnership enables us to carry out the following tasks:

V3 🔁 Reclaiming nutrients from wastewater

The Water Authority Amstel, Gooi en Vecht and the City of Amsterdam are trialing processes to extract more nutrients from wastewater. These trials are carried out at various locations, including multistorey apartment buildings in Buiksloterham. They save 30% of wastewater and generate biogas. We will evaluate these trials, and in 2026 we will determine which initiatives will be scaled. Connecting this with the waste chain, we examine optimal applications of food waste shredders for kitchen waste in high buildings.

This is being demonstrated in Buiksloterham.

Separating organic waste in Amsterdam

We are continuing to implement organic waste collection (including vegetables, fruit, food waste and garden waste) across Amsterdam. The target is to collect organic waste from 30% of Amsterdam's households by 2026 (in 2022 this was just 5%). Our ambition is to extend this to 75%, provided sufficient funds are available. The City also recognises the economic and social added value of neighbourhood initiatives, such as local wormeries and the waste-to-harvest initiative Afval naar Oogst. In local initiatives, those that are directly involved contribute to the circular transition, which is why the City facilitates and supports them.

V4 ⟳ Circular water treatment

In wastewater treatment, Waternet will reduce the impact of the treatment and increase efforts to remove hormones, pharmaceuticals, and microplastics from the water before discharging it to surface water. When treating water for drinking, we already reclaim calcite. By 2026, the Water Authority Amstel, Gooi en Vecht and the City of Amsterdam will have created a roadmap detailing how Amsterdam's water management can become 50% circular by 2030 and 100% by 2050.

Consumer Goods by Value Chain

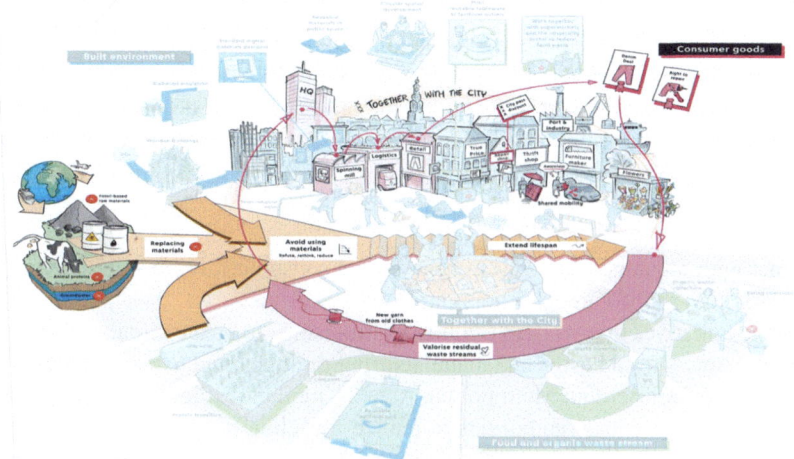

Of all the things we use, textile goods and electrical appliances tend to have the biggest impact on people, the environment, and the climate. As well as mattresses, furniture should also be included here, as this makes up at least half of the bulky waste in Amsterdam.

The 2023 report by the Intergovernmental Panel on Climate Change (IPCC) shows that 40–70% of CO_2 emissions can be reduced by lifestyle and behavior changes. Low levels of engagement mean that consumer behavior is determined by impulses and emotions (ease, attractiveness). Omnipresent advertising encourages consumption. Buying something new is often not just faster, but also easier and cheaper than having a product repaired. As a municipality, we need to think about what we can change in the city to encourage circular behavior.

For these reasons, the City aims to:

• Provide more space for circular businesses, such as repair shops and lending services for commodities, to offer their services, for example by establishing a hub for circular crafts and makers

- Encourage businesses to provide information about the impact of their products and how to extend their longevity
- Consider limiting advertising in public space for "harmful" products—a similar project is being prepared in Haarlem
- Ensure greater transparency across the chain: where do raw materials come from, where do discarded textiles end up, and what impact does this have on the environment globally?

Hereafter we list additional actions for **specific product groups**:

C1 Textiles

Of all consumer goods, textiles are a focal point due to their high impact.

The production of a single pair of jean uses between 10,000 and 25,000 L of water. Due to a variety of factors, including the presence of the design departments and headquarters of various textile businesses in Amsterdam, there is sufficient scale in the region to realize a wide range of partnerships across the chain. This means we can:

- Partner with retailers to raise awareness among consumers, for example through a new edition of the Circular Shopping Route
- Support initiatives that work to establish partnerships to create circularity in the textile chains—from design and manufacturing through retail to use and reuse—as part of the Amsterdam Metropolitan Area's Circular Textiles Green Deal
- Partner with other local authorities to realize recycling of old textiles to create new yarns on an industrial scale
- Trial new types of PPE in a hospital to reduce microplastics in the water
- Utilize our own legal responsibility by maximizing the processing contracts of collected textiles in order to stimulate the creation of circular chains in consultation with manufacturers and other partners in the chain

C2 City Pass discount for repairs

Following the success of the City Pass (Stadspas) discount for garment repair, we will also offer discounts for appliance repair. If there is sufficient interest, we will also offer a discount on the purchase and/or hire of reusable nappies.

Garment repair with City Pass discount

City Pass holders can get a 40% discount on clothing repairs. This initiative ensures clothing is used for longer, helps people with limited means, and provides an impulse for local employment.

C3 🔁 Cutlery and tableware for fast food and event catering

Since the ban on several types of single-use plastics (SUPs), we have noticed that plastic-free alternatives, such as straws and cutlery made from paper or bamboo, are ending up as litter. The transition to reusable products is lagging. To further limit littering and material usage, the City will work with pioneers in the event and hospitality industries to develop trials for reducing non-plastic single-use packaging. We will research which legislation can be leveraged for us to act more decisively in this regard.

C4 🔁 Reverse logistics

The City will enter into new agreements with the packaging industry and with mattress manufacturers regarding reverse logistics and mandatory acceptance of returned products and research how the City can accelerate these processes. The reason we are beginning with these two sectors is that in national legislation they fall under Extended Producer Responsibility (EPR). Reverse logistics is also crucial in other sectors, and we are also examining how we can facilitate circularity in these sectors in Amsterdam.

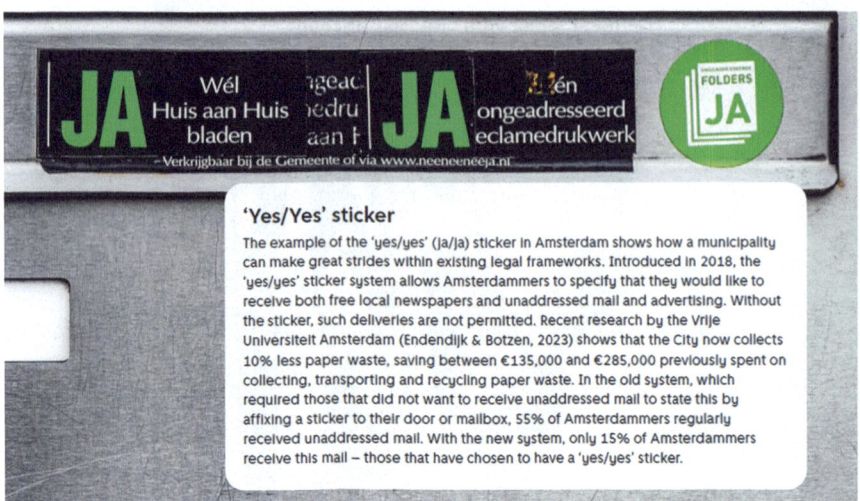

'Yes/Yes' sticker

The example of the 'yes/yes' (ja/ja) sticker in Amsterdam shows how a municipality can make great strides within existing legal frameworks. Introduced in 2018, the 'yes/yes' sticker system allows Amsterdammers to specify that they would like to receive both free local newspapers and unaddressed mail and advertising. Without the sticker, such deliveries are not permitted. Recent research by the Vrije Universiteit Amsterdam (Endendijk & Botzen, 2023) shows that the City now collects 10% less paper waste, saving between €135,000 and €285,000 previously spent on collecting, transporting and recycling paper waste. In the old system, which required those that did not want to receive unaddressed mail to state this by affixing a sticker to their door or mailbox, 55% of Amsterdammers regularly received unaddressed mail. With the new system, only 15% of Amsterdammers receive this mail – those that have chosen to have a 'yes/yes' sticker.

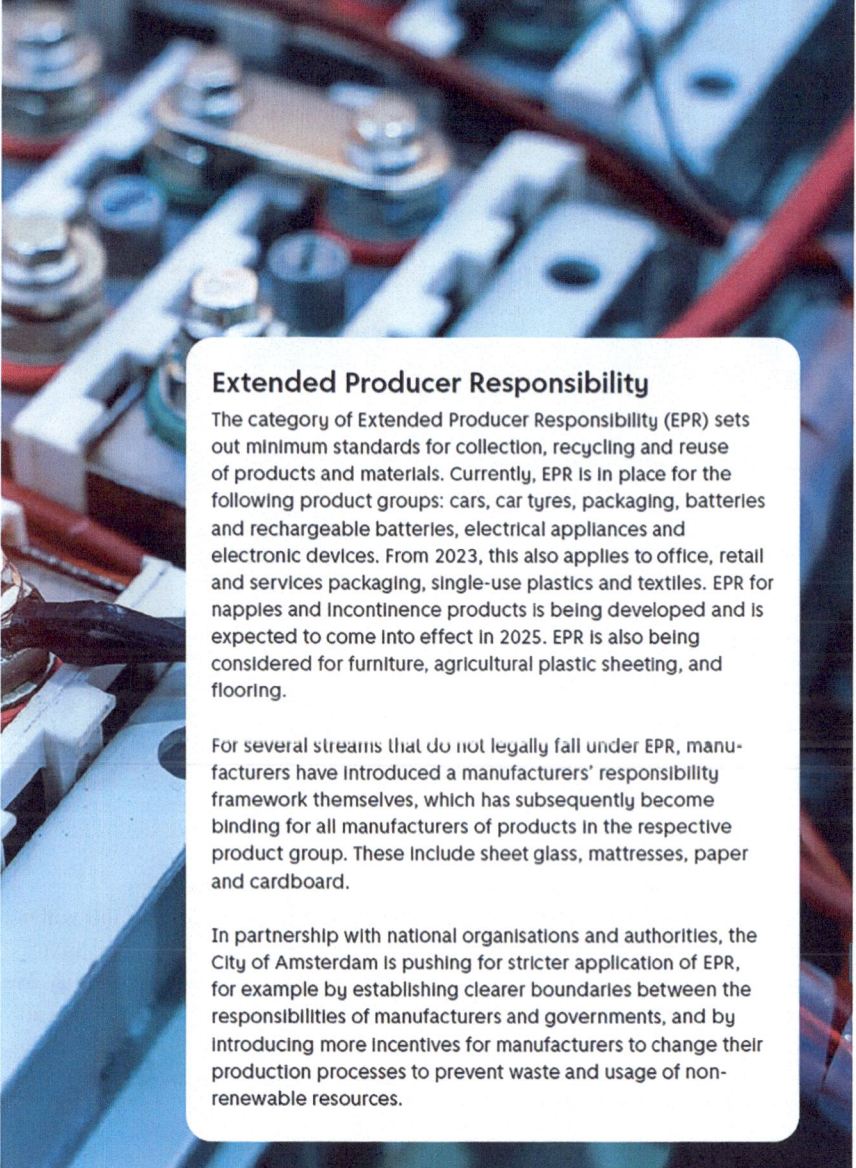

Extended Producer Responsibility

The category of Extended Producer Responsibility (EPR) sets out minimum standards for collection, recycling and reuse of products and materials. Currently, EPR is in place for the following product groups: cars, car tyres, packaging, batteries and rechargeable batteries, electrical appliances and electronic devices. From 2023, this also applies to office, retail and services packaging, single-use plastics and textiles. EPR for nappies and incontinence products is being developed and is expected to come into effect in 2025. EPR is also being considered for furniture, agricultural plastic sheeting, and flooring.

For several streams that do not legally fall under EPR, manufacturers have introduced a manufacturers' responsibility framework themselves, which has subsequently become binding for all manufacturers of products in the respective product group. These include sheet glass, mattresses, paper and cardboard.

In partnership with national organisations and authorities, the City of Amsterdam is pushing for stricter application of EPR, for example by establishing clearer boundaries between the responsibilities of manufacturers and governments, and by introducing more incentives for manufacturers to change their production processes to prevent waste and usage of non-renewable resources.

Preventing waste

In Amsterdam, nearly a kilogram of waste per inhabitant is generated each day. Far too much of this ends up in public litter bins and on the street. In 2023, we will set up a waste prevention plan, in cooperation with businesses and enterprise collectives such as Business Improvement Districts (BIDs). We will prevent waste being generated 'at the front', for example by encouraging a reduction in packaging for takeaway orders, thus preventing environmental impact on public space.

C5 Electrical appliances, electronic devices, and furniture

In 2024, we will create new rules for social thrift shops to inspire reuse, but this will not be enough to ensure a sufficient number of appliances and pieces of furniture are used for longer. We will therefore create more opportunities for this throughout the city. We will produce a guide on where you can have appliances repaired or how you can carry out repairs yourself. We would also like to prevent electronic waste by other means, and we will research the options for this, for example by using the CircuLaw platform.

C6 True price

The City needs to set a good example, and from 2025, we will state the "true price" of our procurement. The idea of true price, or true cost, is important because it creates a more level playing field for circular businesses. It is an extended version of the CO_2 price that is being included in procurement.

The market price of products does not include the negative impact on people and the planet of production, transport, usage, and waste. What we call the true price does include this impact. At present, businesses charging a true price would not be able to compete. This is not because sustainable products are more expensive, but because non-sustainable products are not priced fairly.

C7 ⚡ Fewer new things

The City is setting a good example and buying less new IT hardware and furniture. We make goods such as furniture, laptops, phones, and tablets last longer, for example by opting to repair rather than replace. Ways to make IT more sustainable are being addressed in the Green ICT Action Plan, which was presented to the City Council on April 20, 2023.

C8 ⚡ Circular energy transition

We participate in national working groups that are being organized within the National Circular Economy Program (NPCE) to ensure that the production of solar panels and wind turbine blades is made circular and that reuse of products and reclamation of materials becomes more widespread.

Built Environment

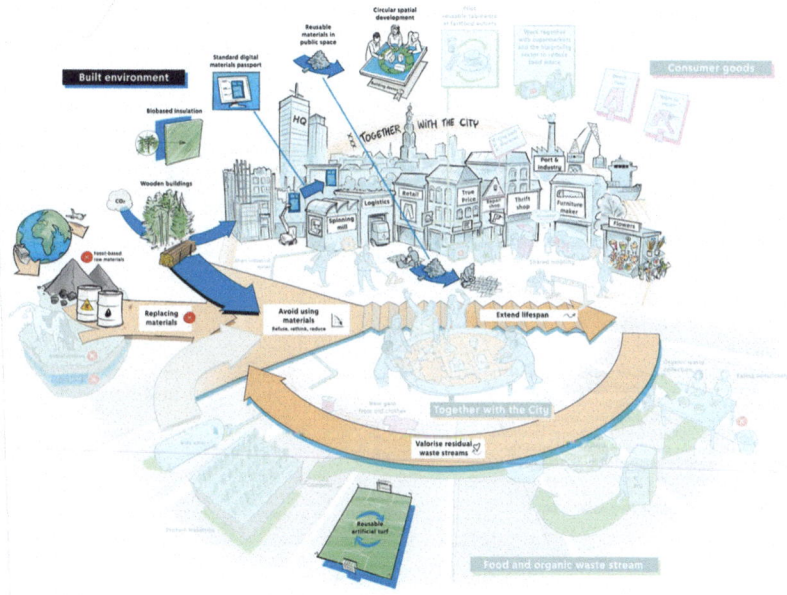

Amsterdam is a compact city. Many people living relatively closely together generally has environmental benefits: commutes are normally limited, public transport is an attractive option, and less space needs to be heated because living space is smaller—in Amsterdam, approximately 12 square meters below the national average.

In addition, a number of measures that are included in the Amsterdam Approach to Public Housing enable a more efficient use of the existing housing stock. We are creating more living spaces within the existing housing stock by facilitating non-traditional forms of sharing accommodation, and we are combating structural vacancy by means of a Vacancy Regulation (Fig. 4.12).

Fig. 4.12 Soil is being temporarily stored by and in the Noorder IJplas until it can be reused elsewhere in the city

Reusing soil

Buiteneiland is IJburg's final new island, and it is exemplary for circular processes. It will be a nature area of approximately 40 hectares, including facilities for sports, recreation and culture. As far as possible, Buiteneiland is being constructed using earth originating from construction and infrastructure projects elsewhere in Amsterdam. Because this material is only becoming available gradually, the construction of the island will ultimately take around 25 years. In other parts of Amsterdam, earth from elsewhere in city is also being reused. For instance, 100,000 cubic metres of soil from the Houthaven neighbourhood was reused in redeveloping Sluisbuurt.

Timber construction in Zuidoost

Between Gooiseweg and Nelson Mandelapark, 700 homes are being built with timber – a great leap towards biobased, circular urban planning.

Actions by Value Chain

At the same time, there is generally a lot of construction taking place, as well as maintenance and renovation of existing homes. The construction sector is responsible for approximately 60% of all material use in the city (calculated in kilograms). This is a value chain in which the City can exert significant influence, not least because approximately 80% of the land in Amsterdam is owned by the City. While partnering with businesses on innovation to find the right balance between circular ambitions and the creation of living space, to continue to reduce the impact of the built environment, the City is working on the following steps:

G1 🔁 Circular insulation

We will help homeowners to insulate their homes using circular methods. Our starting point is that we should find funding to enable them to do this without incurring any costs. Bio-based alternatives should be a full-fledged option that is readily available to all Amsterdammers. We are increasing knowledge about the applications of bio-based construction materials such as cellulose, flax, and cattail, and we are working to provide a scheme to help bridge the existing difference in cost between traditional and sustainable methods.

We are collaborating with organizations from the construction sector and with academic and research institutions to create an overview of which materials are suitable for which applications and situations. We are connecting customers and purchasers with suppliers of bio-based construction materials to promote the development of the market so these materials can become available on a larger scale and be sold at more competitive prices.

G2 🔁 Circular construction

In accordance with the Amsterdam Metropolitan Area's Timber Construction Green Deal, the City will stimulate the usage of bio-based construction materials to reduce the harmful effects of the production of building materials. In spatial planning tenders, we assess circular performance by including the Environmental Performance of Buildings (*Milieuprestatie Gebouwen*, MPG) as a selection criterion. If new innovations lead to the emergence of new, improved selection criteria, we will present these to the municipal administration. When negotiating urban redevelopment projects, we will encourage circularity.

De Wereldburger primary school

A 1960s school building in Amsterdam Nieuw-West has been transformed into a modern, sustainable primary school. It was possible to retain the original concrete structure of the building, while the staircases, doors, ceilings, coat racks and façade were replaced by new versions made from timber and reused construction materials. In the evenings, the school takes on a social function for the neighbourhood. The project was realised by Moke Architecten and the City of Amsterdam on a limited budget. The building was awarded the Amsterdam Architecture Prize 2022.

G3 ☑ Circular urban planning

The City will ensure that all urban planning projects and new designs for area development (including urban regeneration projects and those for public spaces) will conform with circular design principles. The Amsterdam Connected (*Amsterdam Verbonden*) program is working on solutions for urban mobility. The densification of the city which has already been initiated makes an important contribution to reducing the need for mobility, and with it the number of vehicles and number of materials used.

G4 ☑ Circular buildings

The City is setting a good example by entering into a framework agreement for the construction of 30 new buildings, including school buildings, according to circular principles. The first nine circular school buildings will be tendered from 2023.

Community centre Ru Paré

After a turbulent time, the former Ru Paré school in Nieuw-West was converted to a community center by local residents and organizations from the neighborhood in 2016. The makeover was such a success that it won the Amsterdam Architecture Prize in 2019 and the NRP Gulden Feniks. Several years later, the building is now home to a community of dozens of organizations and initiatives, including the De Buurtzaak community center and a theater that offers a stage for art, performance and debates. The building is a good example for how the City can approach its own real estate in new ways.

MK24, foundation for art education, at
Mauritskade 24

G5 🗗 Circular renovation

For renovations and maintenance of its own real estate, the City is using the materials that are already present and/or new organic or reused construction material, as it did for MK24, a foundation for art education at Mauritskade 24.

G6 🗗 Reuse in public space

The City applies the "reuse, unless" principle to material use in public space. To improve the capacities for managing materials, we install new IT systems and create additional storage space.

G7  From country to city

Alongside central government and the region, the City of Amsterdam will partner with construction firms and farmers to produce more bio-based construction materials to stimulate circular construction.

G8 Circular assets: starting with artificial turf

Following a successful trial for sports grounds with circular artificial turf, we will tender the framework agreement for all sports grounds in Amsterdam according to circular principles. The lessons learned will be widely shared, so that we can use them in other procurement processes for all municipal assets.

Bamboo traffic signs

The City applies also circular principles to the production of traffic signs. As a first step, we examine which signs are superfluous and can thus be removed and reused. In addition, we are conducting a trial with the replacement of metal with biobased materials. The manufacture of metal signs uses a lot of energy and therefore causes CO_2 emissions. Bamboo grows fast and captures CO_2. Since 2019, the City has installed approximately 650 bamboo traffic signs to assess their viability in practice.

Artificial turf on sports grounds

Traditional artificial turf on sports grounds has an extremely negative impact on the environment. It is made from new, non-renewable material, releases microplastics into the environment, creates huge amounts of waste and is very rarely recycled. During usage, it heats up the grounds in the summer (to temperatures that can exceed 70°C), and it does not contribute to the intelligent management of rainwater.

What we require is sports grounds that can be used intensively and at all times, without having any negative impact on their surroundings or their users. A successful collaboration with businesses is resulting in artificial turf made from biobased materials. This stays cooler, because it diverts heat to be used elsewhere. In addition, grounds with this type of artificial turf include intelligent water storage for reuse. Finally, at the end of its lifecycle, this turf can be either completely recycled or partly reused.

4.2.5   Ch4: Cross-Cutting Themes, Preconditions

The transition to a circular economy in Amsterdam calls for new forms of organization and collaboration within the municipality. The circular economy disrupts traditional structures and administrative subjects, and this creates obstacles. For now, the City's circular ambitions are co-existing with the traditional way of working, which has been established within a linear economy. To remove these obstacles, there is a lot of work to be done, much of which is not directly visible. This includes the way in which the City is able to conduct its procurement, tenders, design, finance, and management according to circular principles.

New standards are needed for circular products and processes, procurement criteria must be redefined, digital standards set, indicators of circularity monitored, and legal obstacles removed. To achieve all this, the City is working intensively with organizations, companies, European, national and local authorities, institutes such as the Bureau for Economic Policy Analysis (CPB), the Environmental Assessment Agency (PBL), Statistics Netherlands (CBS), the Netherlands Organization for Applied Scientific Research (TNO), and universities.

R1 ⚡ Developing a monitoring system

In partnership with national institutes such as CBS and PBL, we are increasing monitoring to enable us to gain a detailed overview at local level of the various chains and product groups of material streams within a nationally or internationally uniform framework. Equipped with this information, we will be able to work more efficiently on making the economy circular.

R2 ⚡ Developing circular norms and digital standards

Standards create clarity for markets. The City will work with universities, research institutes, local authorities, and other involved parties to contribute to the development of some of these standards, including for the Circular Economy Monitor, material passports for buildings, and various criteria for public tenders. One goal is that these standards are not set solely by established businesses.

We will work with the construction industry and central government to establish a clear definition of circular construction, thus clarifying the criteria the construction sector will eventually need to fulfill.

Digital product passport

The European Commission is planning to make a digital product passport mandatory for consumer electronics, clothes and batteries from 2026. This passport will include all information about the product's entire lifecycle, from raw material to end-of-life, enabling consumer or businesses to see where a product comes from and what materials it contains. In the Netherlands, Platform CB'23 is working on a standard to make a record of the materials of buildings and assets in public space. If it is clear what materials are used in a building or appliance, it will be easier to prolong its life at a later stage or to reclaim materials after the use phase.

R3 ⚡ Sharing Knowledge

We maintain and contribute to expertise and knowledge platforms in the region, the country, and, wherever possible, internationally. This includes information that can be exchanged about possibilities within current legislation, such as the CircuLaw platform, as well as knowledge that still needs to be gained, such as that relating to lobbying for new legislation.

The City has a responsibility toward countries that suffer from negative effects of material use in Amsterdam. We will prioritize those countries when it comes to offering support by sharing knowledge and insights, for example about mitigating measures.

In a regional context, we will connect insights about the circular economy to the programs dedicated to the energy transition to achieve more synergy.

R4 Attracting external funding streams

To finance the transition to the circular economy, it is essential to provide for additional funding. We dedicate capacities to attracting financing from central government and the EU.

R5 Focus on legislation

With CircuLaw, we help legislators make more and better use of the possibilities offered by the existing legal framework. This facilitates municipalities in setting standards and accelerating the circular transition. In the coming years, we will create opportunities to establish measures that are more easily enforceable. This may include:

- Stricter requirements for events
- Increased efforts to prevent food waste in hospitality and retail
- More obligations for reverse logistics, repairs, and processing of appliances by manufacturers and/or retailers
- Stricter rules for public space advertising for products with high environmental costs (comparable to measures currently being prepared in Haarlem)
- Incentivizing or enforcing multiple usage of space by subsidized organizations, for example by making office space available in the evenings for other initiatives.

Regarding ideas for new legislation, we will work with other governments to exchange insights and experiences. We provide input on proposals from the EU or national government. See the appendix "Explanation of activities" for more information.

R6 Addressing the issue of public space

The circular economy requires a reorganization of production chains and other business models. This will create changes in how we use public space and in transport movements. The City will join national and international partners in mapping out this shift in public space usage to be able to adapt spatial planning accordingly.

R7 Continuous learning

The circular economy requires all municipal organizations to change. We are therefore working to create an organization that is constantly learning and that records and shares new insights and experiences. Within the Amsterdam Circular community, connections are being made and differences in knowledge, expertise, business language or jargon, culture, ways of working, and interests are being bridged.

Funding

To implement the activities described in this agenda, the Amsterdam coalition agreement 2022–2026 from June 1, 2022 has reserved funding for the circular economy and sustainability. For the current period, €17.5 million has been made available. In addition, directors of municipal departments have scope to instigate policy changes within the current budget.

For the present agenda, the available budget allows for the following expenditures:

Program	CE 2023	CE 2024	CE 2025	CE 2026	Total
Spring Memorandum 2023	Budget 2024				
CE businesses/SMEs	€550,000	€1,950,000	€2,200,000	€2,100,000	€6,700,000
CE port and industry		€500,000	€1,000,000	€1,000,000	€2,500,000
CE support of social initiatives	€125,000	€800,000	€2,100,000	€2,100,000	€5,125,000
Total "Together with the City"	**€675,000**	**€3,250,000**	**€5,300,000**	**€5,200,000**	**€14,325,000**
CE food and organic waste streams value chain	€25,000	€380,000	€395,000	€420,000	€1,220,000
CE consumer goods value chain		€50,000	€50,000	€50,000	€150,000
CE built environment value chain	€300,000	€230,000	€375,000	€100,000	€1,005,000
CE working on preconditions	€200,000	€200,000	€200,000	€100,000	€700,000
Total value chains and preconditions	**€525,000**	**€860,000**	**€1,020,000**	**€670,000**	**€3,175,000**
Total general	**€1,200,000**	**€4,110,000**	**€6,320,000**	**€5,870,000**	**€17,500,000**

If required, an update of this expenditure proposal will be presented to the municipal administration in the Spring Memorandum of 2024 and/or 2025.

Authors (Gitte Haar) comment on Amsterdam

The Amsterdam case shows how comprehensive it is to drive a transition to a circular economy and how all stakeholders need in depth involvement to succeed. It is impressive how the municipality of a large city in Europe has taken the lead on the transition to a green and circular economy and how detailed the approach has been to engage and implement. The case description here with the four chapters describing: i) the goals, concept and status of the use of materials, ii) the engagement of stakeholders, iii) the value chains and iv) the cross-cutting themes, is very impressive and should be used as an approach to be adapted by other cities and countries.

The approach illustrated in this case very much shows the need to build competences and engage all stakeholders in and around a city. The depth and detailed way of engaging has proven necessary and the City of Amsterdam and their engagement in the future must be a case to follow.

The manufacturer's authorised representative in the EU is Springer
Nature Customer Service Centre GmbH, Europaplatz 3, 69115 Heidelberg,
Germany. If you have any concerns regarding our products, please
contact ProductSafety@springernature.com

Printed and bound by CPI Group (UK) Ltd, Croydon, CR0 4YY
29/04/2026
02099543-0005